HARNESSING TECHNOLOGY
The Management of Technology for the Nontechnologist

D1472561

HARNESSING TECHNOLOGY

The Management of Technology for the Nontechnologist

Stanley L. Robinson

VNR VAN NOSTRAND REINHOLD
New York

Copyright © 1991 Van Nostrand Reinhold

Library of Congress Catalog Card Number 90-24733
ISBN 0-442-00753-1

Manufactured in the United States of America

Published by Van Nostrand Reinhold
115 Fifth Avenue
New York, New York 10003

Chapman and Hall
2-6 Boundary Row
London, SE 1 8HN

Thomas Nelson Australia
102 Dodds Street
South Melbourne 3205
Victoria, Australia

Nelson Canada
1120 Birchmount Road
Scarborough, Ontario M1K 5G4, Canada

16 15 14 13 12 11 10 9 8 7 6 5 4 3 2 1

Library of Congress Cataloging-in-Publication Data

Robinson, Stanley L.
    Harnessing technology: The Management of Technology for the Nontechnologist. Stan-
ley L. Robinson
        p. cm.
    Includes bibliography references and index.
    ISBN 0-442-00753-1
    1. Production management.   2. Manufacturing processes—Automation.
    I. Title.
    TS155.R594   1991
    658.5—dc20   90-24733
                                                        CIP

Contents

Preface

Why is the application of automation on the factory floor moving at a snail's pace? All the evidence points toward an overwhelming reluctance on the part of the American manufacturer to accept with open arms the offerings of the Automation Age. The predominant root causes are the lack of a clear understanding of the needed leadership role as well as confusion over the intricate nature of the technology movement. This is universally true, particularly for the nontechnologian. We are a market-driven economy with a large portion of our industrial leadership by discipline falling into the nonscientific category. Therefore, it becomes critically important to record for posterity the insights and methodologies needed to teach people from all walks of life to manage this new dimension.

The leader does not need to be an expert in computer integration or flexible manufacturing systems. It is more important that he know how to manage the expert. This is the fundamental thesis. The primary objective is to establish technology management for managers from all disciplines that will enable them to step forward and achieve a true and everlasting competitive edge.

The management of technology (MOT) is both top-down and bottom-up. We should expect leadership and direction from the top while encouraging a bottom-up view that clearly recognizes a safe and supportive environment for the idea that automation is an important business advantage. The installation of advanced manufacturing technology (AMT) will not occur until these conditions are met.

The futuristic aspect of this movement tends to create obstacles at all levels of the organization. Dealing with the unknown and the uncertain continues to be a psychological deterrent as it has in the past. Unfortunately, the human being has not changed while his surroundings are rapidly becoming socially unrecognizable and scientifically incomprehensible. It will take a special kind of leadership role to spearhead the proper attack to overcome this adversity.

Upon focusing on the engineer and others who have the direct responsibility for the implementation of AMT, we see another kind of problem. It is no longer possible to stay current with the exponentially increasing Automation Age without taking a special personal initiative. If top-down support is not obvious, this special effort will never take place. Additionally, putting out the day-to-day fires is a burden which leaves very little energy, time, or initiative for planning technology development. Until all these issues are solved, the expected implementation of automation continues at a snail's pace. The solutions center around special nurturing and a unique futuristic pattern of leadership.

Long-term economic growth is the overriding supreme objective. In order to achieve this goal, the manager of technology must establish world class manufacturing (WCM) as a compatible issue. However, the vehicle that will enable us to attain this desirable state is advanced manufacturing technology. Simply said, AMT will lead you to WCM, which is necessary to achieve long-term economic growth. The application of AMT to each business or to an individual plant is the key element in the formula.

Automation is, among other things, a business. It depends on vendors selling automation products to users. If this interaction does not successfully occur, the entire movement is stymied. It is paramount that the user learn how to seek out, assess, and apply AMT. This can only come about through knowledge and an unimpaired relationship with the vendor. Unfortunately, the contrary is true and the relationship is estranged. Consequently, both parties need to change their behavior patterns. However, the user manager properly armed with correct strategy and leadership can go a long way toward promoting a friendly business atmosphere.

The Automation Age is moving so rapidly that it is outstripping reason and overlooking the necessary fundamentals. Healthy evolution needs time for all the proper pieces to fall into place so that the outcome is rational. Herein lies a major cause of the rejection and confusion impeding the progress of this great movement.

In 1941 the attack on Pearl Harbor violently shocked the U.S. into a state of emergency. The sheer reality of a threat to personal security instantaneously catapulted the populace into full recognition of the need for urgency. We were at war and we responded with an overwhelming increase in productivity never before recorded in history. The result was we won the war.

We are now at war. The difference is that the solution does not rest with the military. We are in an economic war which is primarily technology- and competition-driven in nature. Unfortunately, it does not naturally create a realization of urgency among the populace. Instead complacency prevails. There are many reasons for this; to list a few: The percentage of the population employed today is greater than ever. Inflation is reasonably low and under control. The standard of living exceeds all previous levels. There is no scarcity of affordable goods and plastic credit is readily available. Consequently, tomorrow never comes and personal threat does not begin to reach credible proportions.

However, economic warfare works just that way. It infiltrates your environment while refraining from affecting your awareness until it is too late. What are the signs? The most obvious sign, but yet not so obvious, is the trade deficit. The overwhelming situation is our inability to export created by our lack of competitiveness. Unfortunately, this reality is mythical for too many and does not adequately broadcast the need for urgency.

More subtle infiltrators exist. For example, nearly half of the U.S. patents are awarded to foreigners. Over 50% of the Ph.D. candidates and over 20% of the engineering/scientific students in U.S. universities are foreigners. In some U.S. universities, more than 75% of industry-sponsored research is foreign supported. These, as well as other causes, create the final insult in that the manufacturing segment of our economy has fallen significantly. These signs are typical of creeping decadence that classically takes place in economic warfare. On the surface the picture is rosy, but underneath there is cause for concern. Unfortunately, the shock that creates a healthy hysteria does not take place until it is too late. We must address this dilemma by including culture change as an important part of MOT. In addition, a clear understanding of this methodology becomes a critical element of a new leadership profile which we call automation-driven management.

Will we win the war? In all candor, I must admit that I am not sure. History shows that all great empires have fallen through decadence. Are we in the preliminary stages? Can we alter the path? The elements of the Automation Age are all around us. They lie there in indescribable, unrecognizable, almost unusable form in a random mass, somewhat like a quagmire. In fact, we are now in the quagmire of the 90s and a new style of leadership, i.e., automation-driven management, is the real solution. This type of leadership deals head-on with this enigmatic quagmire. In order to untangle this problem, it will take special focus and intelligence blended in a special way. Times have changed, therefore the manager must change. The prime fundamental for automation-driven management is to provide coexistence for automation, the interloper, with all other disciplines in the current state of the organization. The existing organization design has served us well for many decades. It is critical, however, that we accept and execute an expansionary strategy that includes automation in a mode that preserves historical advantages. There is no other way to harness the technological explosion.

HARNESSING
TECHNOLOGY
The Management of Technology for the Nontechnologist

Chapter 1

The Doom Boom

THE DOOM BOOM

The doom boom has been swinging for some time. The issues of competitiveness, automation, technology, and the economy are its prime targets. The realists have taken heed but the perpetual optimist continues to say, "We'll make it, we always have." The trouble is there are too few realists.

There is another school that says, "We'll find a way." Yet we fail to even glimpse that "way." There is surely a lot of activity, discussions, and literature in the environment but the doom boom continues to swing.

Japan bashing has become behavior of poor taste. If you continue to extol the virtues of the Japanese industrial triumph, you are considered insensitive and socially unacceptable. Well, let's switch to Korea and Taiwan, and when they fall out of favor you will have Singapore and Hong Kong. The trouble is there are too few realists and that situation only serves to feed the overwhelming inertia.

It is wrong to be the carrier of doom and gloom, leave your audience in a total quandary, and then move on to the next victim without providing a comprehensible solution. I promise to refrain from leaving you in a depressing vacuum. It is the purpose of this book to specifically lay out the proper way to harness the technological explosion so that the nontechnologist can manage.

However, first we must lay out all the gloom and doom in order to thoroughly understand and define the problem. We will do that in finite terms. Second, and most important, we will spell out a strategy that is comprehensible to the nontechnologist that takes you to the competitive edge. Of course, this strategy will vary by type and size of business. However, there is a central tendency around which one can adjust and tailor the specific situation.

DOOM AND GLOOM

Things are bad depending on who you talk too. There are more people working than ever before. The standard of living is the highest ever for the majority. Inflation is low and under control. Minorities have greater opportunity. The underprivileged can send their children to college. Credit is readily available and affordable goods are overloading the shelves, although all to frequently they are not labeled, "Made in the USA."

So, why worry? Free enterprise continues to prevail. We are still the greatest country in the world. It then follows that our government must be the greatest in the world. So, why worry? It always has been and it will always continue to be. It is also historically true that the same feelings prevailed at one time with the Romans, Kubla Khan, and the British Empire.

In order to define the problem, it is best to divide the situation into four categories:

1. Government

2. Academia

3. Industry

4. Environment

GOVERNMENT

We have the best government in the world, as well as the greatest country in the world. I have served in the U.S. forces in World War II in the Atlantic and Pacific Theatres and hated every minute of it. More important if need be, I would serve again. We do have the greatest country in the world, without question.

Part of that greatness is freedom of speech and you can even choose whether you wish to listen or not. The problem is that too many people are not listening to what's going on. On the other hand, from the point of view of global competition and our government there is very little to listen to. There is considerable government research underway in technology, mostly oriented toward defense. Although some of this is transferable to industrial application, it is not highly publicized.

Even though we do not expect our government to interfere with private enterprise, it would be appropriate to expect a recognizable public leadership profile on the issue of competitiveness, automation, and technology. The best you will find is an overall cloudy umbrella position that says, technology is vital to our future and economic growth. You could write a book on what this does or does not mean.

The Manufacturing Studies Board of The National Research Council (made up of members from all phases of academia, industry, and government) have recently concluded that the problem centers around three major issues:

1. Competitiveness

2. Advanced manufacturing technology (AMT)

3. Management

Beyond this they did not specify a path to follow that would provide a solution. It appears at this stage that there is no clear answer, although this work contributed in a worthwhile manner towards defining the problem.

The three issues simply explained seem to center around the following ideas:

1. Competitiveness on a global basis is unsatisfactory

2. AMT is not understood by too many

3. We need a special style of management to deal with the future

In summary, it appears that the right kind of overt leadership behavior in government is missing. From this vacuum then develops an erroneous aura of disfavor, i.e., that the government is not really sponsoring these issues. The lack of government as a prime mover then places them in an "on the fence" position and, most important, leaves them in a bland plateau of tranquility on a non-vote-getting issue.

We can conclude from this that a national banner will probably never emerge on the issues. Therefore, a strong central thesis will not serve as a catalyst to develop a universal momentum that will catapult the USA to greatness once again. Therefore, since we cannot rely on this expectation, we must find another way. The answer lies with the individual leadership of every business throughout the country. Hence, this book becomes ever so important since its objective is to teach you how to harness the technological explosion.

ACADEMIA

Academia is made up of intellectuals. They stand ready to train managers for the future. The question is, how do you do that? Universities are now searching for the answers and are developing new curricula which will hopefully solve the problem. It is recognized that the implementor of AMT must be the engineer in manufacturing. Therefore, we also see the universities testing new curricula that will equip the engineer of tomorrow to meet the challenge. The problem is that we won't have the answer in a practical way for at least two decades.

It is said that the Automation Age is proliferating new technology at an exponential rate. In fact the pace is presently beyond the absorption rate of the practicing engineer and is one of the problems we will deal with in later chapters. The university, with its proliferating research programs, is a major contributor to this problem, and rightly so, since the alternative is unacceptable.

Centers for AMT, productivity, and manufacturing technology are springing up all over the campuses offering advanced degree studies and an array of excellent research. The problem is we are not using it at the plant site fast enough. Many of the colleges solve this problem by making their findings available to foreign sponsors. Is this wrong? It is compatible with free enterprise and solves a business problem, but on the other hand, it serves to aggravate the problem at home.

In most cases the centers have invited U.S. industries to enter into a partnership. These alliances provide the school with real-life feedback and the sponsoring industry gains a source of knowledge. The universities now cover only a miniscule fraction of the industries in the country and find in general that this proposition is a "hard sell."

It is interesting to note that over half of the Ph.D. students in the U.S. are foreign nationals and an increasing percentage cover our total engineering student ranks. There is clear evidence that we are potentially sharing our knowledge with foreign interests. The problem is that too few of our own high school graduates are qualified by virtue of training or ability to enter into a demanding curriculum such as engineering. Our offshore competition do not have that problem. In order to keep our ivy halls open, academia have no choice but to fill their classrooms with foreign nationals.

Since a large portion of the Ph.D. students are foreign nationals, it follows that a proportionate amount of professors and instructors are also foreign nationals. You could even go as far as to say that not only is U.S. manufacturing out of global competition, but so are our institutions of higher education, since American youth no longer have the "smarts" to compete.

There is more gloom and doom to come—do not despair. So what can we conclude from all of this? First of all, we need an overhaul of the high school system. By the time this is settled, it will be too late. The massive school system bureaucracy plus the special interest groups will see to that. No doubt change will come, but the timeliness of this change is the grave question.

Academia has to remain vital in the meantime. That is critical. We are fortunate that they are a visionary society, but they sorely need better applicants from the ranks of American youth. In the meantime, with feedback from their industrial partners they will also find ways to change the curricula and develop engineers/managers for the future. Unfortunately, not soon enough.

Academic research will continue to escalate; a great deal of it will leave the country; and we will continue to innovate further. That's the nature of the syndrome. This will continue to create confusion, insecurity and ignorance, although we can expect the user of automation to improve his capabilities slowly. However, it is questionable that we will ever win the race by this haphazard method of technology transfer.

How can we change all this? There is only one answer. It lies with the individual leadership of every business throughout the country. Hence, this book becomes ever so important since its objective is to teach you how to harness the technological explosion.

INDUSTRY

Industry would be totally automated today if we could, but we can't. We would be Number 1 in global competition today if we could, but we can't. Our industrial society has evolved into a culture that prevents this from taking place expediently. Let's examine this situation, "if we could, but we can't." What is preventing us from achieving our destiny as the greatest competitor in the world?

In the first place, from the CEO down to the junior engineer, very few people truly understand the nature of the Automation Age. The financial officer, the research director and the manufacturing executive are not prepared to interface with each other on this major issue in the correct way. Even the practicing engineer is not prepared to cope with this futuristic environment. When a cultural linkage is eventually and clearly established between the CEO and the engineer we will at last arrive at the desired frontier. We will deal with this matter in later chapters.

It has often been said that we are a fad culture. A few years ago, productivity was our salvation. That lasted a while, leading us to quality circles. Next on the horizon came other total quality concepts, quickly giving way to material resource planning, followed by just-in-time. None of these fads really faded away completely. They are still around, but they don't carry their former focal point emphasis. What's next?

Well, the truth is we don't have a "next." The issue of AMT is too complex, so it does not easily become a "rally around the flag" issue. AMT is not a fad. It is a serious basic issue of great potential impact that we as a country are not truly prepared to deal with. If it were a light and simple concept, you can rest assured that our entire industrial society would be deeply embroiled. Everyone would try out the idea on their factory floor.

In fact, that is what happened to robots. The first visible, apparently, easily understood aspect of AMT were robots. Everyone rushed to the vendor to get one although they didn't have an exacting application. The attitude was that it was necessary to get into the vogue. The result was that robots were found to be more complex than they appeared. You have to know payload, manipulators, safety rules, robot teaching methods and on and on. The upshot is that the robot industry is not achieving forecasts. Finally, the happy recipients of these follies are the research centers in our universities. I believe that tax reform still allows deductions for contributions to learning institutions.

It is not that there are no examples of "lights out" plants in the USA. Some companies have shining examples of automated factories. There are others who have tried and failed. It takes more than money to be successful in AMT. The problem is that on a percentage basis the shining examples overall are very small. Again the necessary change will not take place until the power structure of our industrial world are ready to share that power with those who can make the difference. The proper signals are not travelling between the various camps in the organization. The message is incomplete.

Does the solution really lie in the sharing of power? Can it be that simple? What is the probability that "them that has will be the ones that give?" The concept is worth careful examination and we will do so in later chapters. It all boils down to the same answer. The individual leadership of every business throughout the country must take the initiative. That is why this book is so important, for it will teach you how to harness the technological explosion.

ENVIRONMENT

Listen carefully to the pulse and vibration of the environment. Over and over again you will hear the same messages piercing the din. We list a few typical ones for discussion.

1. The U.S. is in the longest economic expansion in our history.

2. America must remain competitive as we move into the 1990s.

3. We must improve quality and productivity.

4. The government should provide an environment for entrepreneurship and risk taking.

5. America has the ability to compete with anyone.

6. We must work harder and smarter than our competition abroad.

If you listen to the environment, you will hear these messages coming through from multiple sources. The real message is absent. How do you really go about achieving and perpetuating all of these objectives? The answer still lies in the initiative of the individual leadership of every business throughout the country.

The environment will continue to be abound with these vapid messages. The problem is that too many citizens believe them on face value. You hear it long enough and you assume that it must be true. Let's take a look at each of these messages.

1. The U.S. is in the longest economic expansion in our history.

It is true that we are in the longest economic expansion in our history. Unfortunately, it is slowing down. GNP growth is less in the 'eighties and 'nineties than in the 'seventies. Each year we continue to see some gain, but the overall rate is diminishing. The important question is, where will we be in two or three decades if we don't reverse the trend? More important, what is our planned strategy to make a turn around happen?

2. America must remain competitive as we move into the 1990s.

I'm sure we all agree, and I'm confident that you've heard this said before. How do we go about making this come true? Unfortunately, too many have chosen to avoid this challenge. They have concluded that reducing cost through automation is too risky. There is an easier way. The easy answer is offshore manufacturing. If we can't do it, we'll find somebody who can. We are free in America to make these kinds of choices. Until we see the trade deficit disappear, we have to recognize that we are no longer number one. An expansionary economy is unsound until this blight is removed.

There are those who hope for a further weakening dollar to stimulate exports. The currency exchange is the last basis upon which to plan an economy. That means that you are totally out of control.

Then there are those who favor protectionism, which in the long run will raise concerns about inflation. Reciprocal trade is another possibility. Unfortunately, there is only one safe answer for the dilemma and that is true competitiveness on our own shores.

3. We must improve quality and productivity.

A simplistic statement that rings totally true. If so, what's the big deal. The problem is our culture. Our management culture, labor culture, education culture, social culture, government culture, and on and on. They are all intertwined, inseparable, and end up in an attitude among the populace that truly does not appreciate the great advantage of living in the U.S.A. Why is it necessary to even raise the question of quality and productivity. Under reasonable conditions these two considerations would be ingrained in every man and woman in the workplace.

Our culture may very well end up being a threat to our long-term security. For without some form of shock treatment there is no response. Quality remains a problem and productivity is waning. The culture that has evolved does not naturally rally the citizenry to preserve what is the greatest country in the world and its intrinsic values.

4. The government should provide an environment for entrepreneurship and risk taking.

Unfortunately, the government doesn't know how to do this. On the other hand, their restrainment of interference is in itself a benefit to entrepreneurship and risk taking. It allows the manager freedom to fulfill his role if he truly understands the management of technology. The results to date support the idea that we need a special approach to teach the power structure, which is primarily a nontechnical community, how to deal with MOT, the management of technology, and to harness the explosion that it has created.

5. America has the ability to compete with anyone.

Beyond question this is true. We are the great innovators and still have a freedom-oriented environment that nurtures creativity. But something has happened. We are enshrouded in a blanket of darkness that serves only to restrain our talents in a state of lethargy. The error is in the prevailing doctrine that freedom is free. History provides evidence that wherever this casual attitude has prevailed so has failure. Obviously, the kind of leadership that can make the difference is evading us. It is time to mobilize and reverse the trend. It is time to mobilize and teach our managers the techniques of automation-driven management. Then most assuredly we will compete with anyone.

     6. We must work harder and smarter than our competition abroad.

It has been said in many ways that our competition abroad work harder and smarter than we do. During World War II we proved that we could do that better than anybody given a state of emergency. In the perception of most we are not in a state of emergency today, and, consequently, that extra needed focus does not exist.

If we heard it daily on television, or read it in the newspapers, it might begin to sink in. A controlled hysteria would take place and then we would respond by working harder and smarter. Without the proper impetus the culture of the U.S.A. will not change in the short term. Additionally, if that proper impetus does not appear, and we continue in the current direction, it is doubtful that we will truly work harder and smarter. In order to make the difference, it requires the individual leadership of our businesses to manage in an appropriate way that is compatible with the current automation age.

CONCLUSION

You can see that a major portion of the problem is social as opposed to economic and technical. It is problematic socially because the needed changes must take effect among people first since they are the ones who control.

It has been proven many times that leaders are emulated by their subordinates. Behavior change then will occur on a top-down basis. It is, therefore, important that we provide our leadership with the desired behavior pattern that will promote the special activity of technology management. This is the major objective of this book. We wish to teach MOT for leaders at all levels of the organization in a practical nontechnical way.

With this approach, we are placing the emphasis where it can have the greatest effect. Needless to say, the majority of our industrial leaders are nontechnical in background. So, why not foster MOT for the nontechnologist. There is no other way to achieve progress in the short term. It is the short-term results that will save the day. A state of urgency exists. We must move now, for any further delay brings us closer to failure and makes recovery that much more difficult. Our government is quiet. Academia is willing to help. Industry is floundering, and the environment lacks in public awareness. The individual leaders of our businesses can control their sphere and create that public awareness where it counts. They can stop the Doom Boom from swinging. There is no other way to harness the technological explosion.

Chapter 2

Back to the Basics

THE PROBLEM

Automation is a must. There is no question that manufacturers are under mandate to eventually move to this futuristic mode. World competition has already made this obvious. However, automation should not create such a preoccupying focus that the basics of manufacturing are overlooked. It is unwise to overlay advanced manufacturing technology over an unsatisfactory production foundation. Unfortunately, the underlying deficiencies are perpetuated and adversely affect future performance. You should never automate your mistakes.

Furthermore, there is a substantial segment of the producer community that will probably never reach the status of having a full-fledged advanced technology program. Limited capital as well as other scant resources are the predominant restrictions. It becomes crucial that this group pay particular heed to the basics of manufacturing excellence. It is their most important way of maintaining competitiveness.

One could question whether our world competitors are fundamentally better manufacturers. The fact that we export our manufacturing to Korea and Taiwan supports that idea. They are not highly automated and yet they are extremely competitive, even against Japan. Of course, we immediately salve our conscience by rationalizing that the massive spread in labor rates make up the difference. If the labor rates were equal would the offshore producers still come up the winners?

All the evidence seems to point out that our world competitors work harder and smarter than we do. Has the U.S.A. developed a culture that spawns an inferior manufacturing system? Wherever this is true, then I would offer that a major portion of the fault lies in the fact that we have forgotten the basics. It is time that we take inventory and carefully consider the need to augment our strategies by among other things going back to the basics.

The forces that have carried us to our present undesirable state have been varied. For example, one of the major issues has been the absence of need. It has not been necessary to strive to be the best in the world. The major factories abroad had been destroyed by the allies in World War II. It has taken several decades for the offshore producer to recoup. In the meantime we have been leaders but "asleep at the wheel."

Secondly, the neophyte at the entry level about to enter the industrial scene has already been turned off. The engineering/MBA graduate has decided to seek more prestigious, lucrative fields. The hard-nosed, shirt sleeve production superintendent with an engineering degree has faded away. In his place is a computer-oriented strategic thinker who does not really know what's happening on the production floor. In fact he has never been properly exposed to the culture that exists there, tends to leave too many of the details to lower level front-line personnel, and never achieves a conversational ability in their occupational language. Consequently, he settles for a lesser result in his factory since his judgment is dampened by lack of exposure. On-the-job, first-hand manufacturing experience is not available in the ivory tower.

THE MANUFACTURING IMAGE

It has been said that manufacturing as well as engineering personnel are no longer first class citizens. Business is market-driven. Even though the operations group are essential, they are not totally accepted in the board room inner sanctum. Who is too blame? Both parties, of course. The market driver is too quick to subordinate the value of manufacturing excellence, while the factory managers are all too ready to accept the "surprise packages" "thrown over the wall" by the other disciplines. Should manufacturing continue to be a reactive service function to the rest of the organization? Assuredly, this image is not in our best interest.

The topic of design for manufacturability (DFM) is treated as though it is a recent invention of the Automation Age. DFM is defined as placing product and process development in a unified effort. Unfortunately, this is one of the basics that has fallen by the wayside. It is its existing notable absence that makes DFM appear as an innovation. All too often the production group is the last to know what must be produced. The imbalance of power in the organization, as well as the lack of two-way communication, has allowed an unbalanced equation that assigns the production group to the lesser side.

The organization leader who has the power to balance this equation hesitates to do so because of his concern for the monthly P/L statement. Instead his overwhelming emphasis on sales optimization creates a view that manufacturing is expected to stand ready to respond to the whims of the power structure. A quick look at the past shows this to be true and by and large manufacturing has reacted in exemplary fashion.

Mahogany row leadership has evolved as our overwhelming business culture, conceptually placing the plant a million miles away. Knowledge of manufacturing results are viewed as an arm's-length issue. Top-down focus on these issues may occur only as the monthly P/L statement is circulated. At this point it is too late and if profits are at an acceptable level then scrutiny of the cost of goods sold may never take place. The ivory tower executive has more important matters on his plate.

THE BASICS

All too often there exists even among the operations personnel a distortion as to what is important. It is the objective of manufacturing management to produce a high quality, low cost product in a timely and safe manner. Quality and safety should always be in an override status. This means that these features exist in perpetuity and are never sacrificed or subordinated. All activities are carried out against a backdrop of primary consideration for quality and safety. Now let's look at the remaining aspects of the basics of manufacturing excellence.

Generally speaking the cost of goods breaks down into 80% for material, 15% for overhead and 5% for labor. Consequently, materials management and particularly the purchase price of parts and raw materials become paramount issues. But what about material utilization, is it not equally important. In spite of this reality waste control is left up to the front line personnel and any further attention is diluted by the nature of P/L reporting.

Furthermore, acceptable waste allowances have become an accounting football. There is too much concern over "looking good" as opposed to taking a stringent point of view as to what can really be achieved. Operator training, maintenance management, supplier reliability, and machinery modification are all tools that are related to materials control but are put aside for more lofty matters. The manufacturing mahogany row strategic thinker rarely looks in on these matters if at all. He delegates to those in the trenches.

Machine efficiency is also an important relevant factor of manufacturing. In this case machine efficiency would be defined as follows. If the process capability is 1000 units/hr, then a yield of 900 units/hr would equal 90% efficiency for each hour of scheduled time. In this example we are concerned only with the machine utilization of the time the process is scheduled to run. It then follows that minimal downtime or continuous running time means higher efficiencies yielding more product. This is the life's blood of manufacturing. Also continuous running time without interruptions yields less scrap and more acceptable product. In addition, machine efficiency is a relative measure of both overhead and labor utilization as well. Consequently, machine efficiency is without doubt a critical measure of the production system.

My experience dictates that productivity is also an important relevant factor. The optimal utilization of inputs to yield maximum outputs is indeed an important measure. However, for day-to-day factory floor consideration I would recommend a simplified version, as follows. For example, if 1000 units/hr is required to meet the established goal then 1100 units/hr would equal 110% productivity. At this level of output the manufacturing system is presenting an opportunity for the business to make full profit on each product sold.

The number of people on the job or head count is the fourth important factor. If the specification requires X workers and your daily report shows $X + Y$, then it is important to find out why. Granted labor is only 5% of the pie. However, manpower overage usually indicates a problem which probably contributes to losses greater than just the cost of additional head count.

The nucleus of manufacturing excellence, given the quality and safety override is:

1. Low waste

2. High machine efficiency

3. High productivity

4. Head count

THE BASE OF EXCELLENCE

In fact these are the primary measures of the manufacturing team. This is where it all begins to happen. This is where we establish the manufacturing base of excellence. It has to happen at this level to avoid decadence at more complex levels. Therefore, the first thing each morning all production personnel should be made aware of these results for the previous day. Before any other issues should be taken on, the four nucleus values should be reviewed, and, where deficiencies exist, follow up with immediate attention on the factory floor. Then the message is clearly established that management cares.

If the top level manufacturing personnel are seen in the shop asking about these results, then it is clear to all what is really important. In addition the idea that "management cares" comes across, it is infectious, then everybody cares, even the sweeper.

The case has been developed that identifies the nucleus of manufacturing excellence. It is dangerous to assume that this is all that counts. Product design, process development, and many other matters are critically important. However, the ill that is created is that these outer layer issues overshadows the nucleus to the point where manufacturing management fails to monitor the true daily results. The practice of "nucleus first" before all other matters should prevail. Figure 2-1 depicts what I have described as the manufacturing ecology.

You will note that the inner nucleus is made up of the four relevant factors. The next layer exemplifies where many of the managers hide, never allowing their sphere of attention to penetrate this nu-

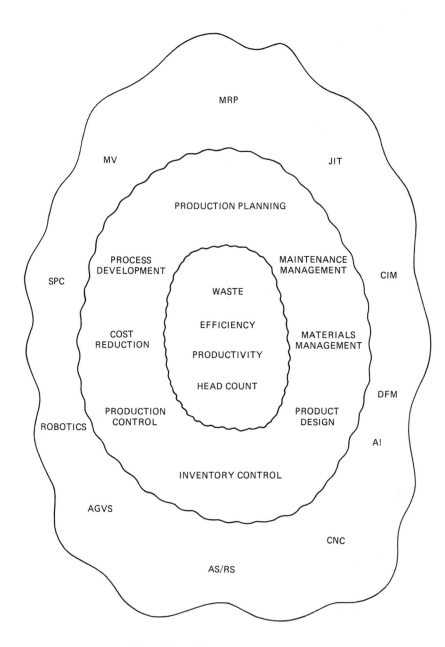

Figure 2-1 Manufacturing Ecology

cleus. In some cases they even contain themselves in this middle layer in order to avoid escalation to AMT issues. The middle layer can be a haven providing two-directional avoidance. You can avoid the nucleus of four relevant factors where you have to roll up your sleeves and get your fingernails dirty. You can also avoid the outer layer of AMT which is perceived as complex and risky.

True excellence in manufacturing cannot be achieved without conscious activity in the nucleus. Additionally, it is erroneous to develop a strategy for expanding to AMT before the four basics are stabilized in a state of excellence. This statement is based on recognition of the quality and safety override while also recognizing the importance of the outer layers of the manufacturing ecology. Unfortunately, this is where U.S. manufacturing has given up their leadership in global competition.

OPERATIONS STRATEGIC PLANNING

Now we shall talk about the glue that maintains a bond throughout the manufacturing ecology. It is operations strategic planning. In fact one of the major criticisms of U.S. manufacturing is that all too often there is a vacuum where an operations strategic plan (OSP) should exist.

The OSP is developed in compatibility with the overall business plan. It spells out what operations objectives must be achieved in order to meet the overall business objectives.

It serves a number of purposes. First, you know in advance what has to be achieved as opposed to catching whatever is "thrown over the wall." Second, you become automatically involved as a partner with the other disciplines. And, last, you develop a planning base that at the appropriate time can include such futuristic issues as AMT.

The place to start, however, is with the four basics. You have to decide where you want to be in waste levels, efficiency, etc., while establishing programs and projects that will carry you there. By making the four basics part of your OSP, you have reasonable assurance that you will truly return to the basics while establishing the proper manufacturing underpinning for expansion to greater heights. There is an old Chinese proverb that appropriately expresses the true value of operations strategic planning: "If you don't change your course, you will end up where you are heading."

At this point it seems desirable to outline briefly a four basics program. It is possible to elaborate at a sophisticated level on this program; instead, we will talk to the simplest form and allow the reader to expand the program according to his needs.

LOW WASTE

The hue and cry is usually that tracking scrap on a daily basis is a horrendous task. So, let's keep it simple. During the shift you insert raw material into the process at the input end. This should be recorded. At the end of the shift you know your production which can be converted quickly into raw material used successfully. The difference between input and output is scrap.

There may be a partial unit of raw material at the input end of the process at the close of the shift. This should be factored and credited accordingly. You can create simple formulas for this calculation. Remember that you are looking for a daily barometer and 100% accuracy is not essential.

If you employ this type of approach, then you can work within your individual special conditions. For example, some processes make inherent waste. If so, then this unavoidable fixed portion should also enter in the formula as a credit.

You now have the framework. Once again, it is important to keep the system simple. Additionally, the "scene of the crime" should be visited daily to discuss cause and effect with those who can make a difference, the people in the trenches.

It may be necessary to establish industrial engineering studies to assist in developing solutions. Machinery modifications or supplier reviews may show up as potential activities. It is amazing what daily focus can turn up. Certainly, this approach communicates the idea that "somebody cares."

MACHINE EFFICIENCY

Machine efficiency should be measured during scheduled time. If the capability of the machine while running is 1000 units/hr, then a yield of 900 units/hr translates into 90% efficiency. If downtime (DT) is excessive, then it requires a person with a stopwatch to record each increment of downtime and its cause. So often I am told that this re-cording is done routinely by the personnel on the shift. I can tell you from experience that this method becomes overly casual in time and yields inaccurate interpretation of causes and effects. Responsible people should be assigned to analyze these matters and provide solutions.

The events of downtime can be categorized as avoidable and un-avoidable. Then projects should be established to eliminate the avoid-able items. It is also possible by changing some conditions to convert some of unavoidable to avoidable status. Remember that in this case efficiency is measured only during scheduled time. The people on the factory floor can directly control downtime during this period. Idle capacity is not within their control.

The problem is that our culture of the day shuns "nitty gritty" as described above. I can only say this is where our global competition shines. It is time to concern ourselves with these details and come out of hiding from the middle layer of the manufacturing ecology.

PRODUCTIVITY

For day-to-day factory floor consideration I would recommend a sim-ple version of productivity. Multifactor productivity measures the uti-lization of a number of inputs such as man, material, machines, meth-ods, and money. You could include energy and such assets as buildings as well. The expected value of cost of goods sold is a combi-nation of all these inputs as well. Therefore, this value can be used as a simple measure of the relative multifactor productivity for operations.

For example, if 1000 units/hr are the required production to meet this goal, then 1100 units/hr would yield 110% productivity. Obviously, all of this information is readily available. Needless to say, at a productivity level of 100% or better, the business has an opportu-nity to make full profit at the operating level.

HEAD COUNT

This information is available in the production specification. A com-parison of actual versus specified on a daily basis highlights trouble spots. Usually, extra personnel are used when the process runs into trouble.

Once again, it is paramount that these statistics be reviewed by exception on the factory floor on a daily basis. The entire operations management team should participate in this process in some manner. If the manufacturing system has several locations, then each site should have an individual review process and knowledge of results should be made known by an appropriate vehicle at headquarters.

Of course, these procedures outline the fundamental approach. They will obviously need to be adapted and tailored to fit your own conditions. So often I have been asked at what level should the goals be set. The answer is that at the outset, the goals should reflect expected reasonable short-term improvement, perhaps 5–10%. It has been proven on a longer-term basis that extraordinary results will be achieved by this type of scrutiny. In fact, I have been personally involved in situations where the starting condition was devastating but the four basics ended up within one year as follows:

1. Waste   Less than 1%

2. Efficiency   96%

3. Productivity   110%

4. Head count   Zero variance

CONCLUSION

These are the four basics of manufacturing excellence, i.e., the manufacturing nucleus. Herein lies the base that must be established before considering AMT. In fact, as you move into computerization of the manufacturing process, all these performance measuring systems lend themselves readily to electronic data processing.

Once this type of attention to detail becomes the standard approach for the manufacturing team, then a whole new vista of potential improvements will become evident. In fact performance measurement at a higher level of sophistication can overlay the four basics, yielding broader solutions. The problem is that our current culture fosters superficiality in manufacturing matters. It has been said many times that our global competition attend to details in manufacturing with thoroughness beyond expectation.

Well, we can change our culture by starting with the four basics. It takes leadership with an awareness for this imperative to assure that we truly value and carry out this approach. For without that class of leadership, we will never truly alter existing cultures.

It is time to leave the executive suite, invade the factory floor, and promote the policies that will improve the economic and managerial performance of the firm. The leader must demonstrate that he is a sincere advocate of change and appropriate competencies among them being the principle of back to the basics.

We can expect that leadership presence on the factory floor will be seen by both management and labor as positive. They will rally to the cause given the proper two-way communications as to what is happening. It has been said many times that people can make the difference between excellence and mediocrity, particularly if management cares.

The discontinuities and imbalances that existed yielding less than desirable results will fade away when the "attention to manufacturing details" attitude prevails in the factory at large. This is the heart of the matter and will motivate all of the players, both labor and management to maintain a cooperative problem-solving mode. This is the beginning of the strategic defense, i.e., back to the basics, that will spread out into a labyrinth of inquisitiveness that will truly scrutinize all aspects of the production system and prepare us to effectively deal with AMT in the near term.

Chapter 3

What Is AMT?

AMT means advanced manufacturing technology.

AMT is a major contributor to achieving world class manufacturing (WCM) status.

A world class manufacturer is one who achieves the competitive edge.

Therefore, an important link between WCM and the competitive edge is AMT.

The latter statement contains three important elements, i.e., AMT, WCM, and the competitive edge. The subject of AMT and several related features will be discussed throughout this chapter. What can we say about WCM? Of course, this varies according to the nature of the business. However, as a general uncomplicated overall definition, we suggest the following.

WCM is a production system that yields under safe practices the highest quality, lowest cost, and most timely product on a global basis, while possessing internal management systems that are geared toward perpetuating this leadership position.

The third element is the competitive edge. First of all, we should not confuse the competitive edge with the "leading" or "cutting" edge. Both of these descriptions can take on very narrow meanings, such as a "leading edge" machine or a "cutting" edge technology. The competitive edge in its broadest sense yields the highest quality, least cost, and most timely product in a safe manner. It must also include the most innovative product as well. Therefore, the way to link WCM and the competitive edge will follow a simple scenario.

You first purchase a leading edge process. This is the best possible "off-the-shelf" public system available, and it is also known to your competition. Next, you examine the array of AMT and apply those that are appropriate to this cutting edge process, adding proprietary advantages that go way beyond the conventional. The overlay of AMT on the leading edge process should be designed to clearly provide you with a safe system yielding the highest quality, lowest cost, most timely, and most innovative product in your sphere of competition.

It is also important to remember that the competitive edge is a constantly moving target. Therefore, manufacturing research should provide for a continued effort, along with other management systems that preserve and maintain this leadership position.

It is true that the so-called competitive edge is not dependent on manufacturing alone. A business in totality is dependent on the way all disciplines are managed. However, the popular belief is that manufacturing is the focal point for the future. It is where quantum leaps in change can be achieved! Manufacturing is the window of opportunity, it is said. Without attempting to dispute this issue, we know that the production system is not isolated from the rest of the organization, and its impact on the business is significant. Therefore, it is critical to understand how AMT can contribute to the overall strategic objective.

Additionally, the literature seems to support the notion that our biggest failure to keep pace with global competition has been in manufacturing. There is further implication that marketing, finance, and research and development, etc., are not to blame. Can it be that manufacturing alone is truly the cause of our failure?

Logically, it is improper to indict the production system as the only culprit. What about American management in general? The components of the organization are not discrete. Obviously, the way we manage in total is critical in relation to the complex nature of technology. There is no question that we must augment our management profile with a new dimension that enables the nontechnologist to manage technology. Here, too, is an important reason why an explanation of AMT is fundamental to our ability to achieve WCM. So within this framework, let us take a closer look at the question, what is AMT?

AMT is in reality an important part of the overall automation issue. In fact, AMT is the latest phase of automation and represents its highest order to date. Automation in several forms has been around a long time, and there have been many influences in history that have brought us to our present state. Consequently, it behooves us to take a look at the past two hundred years to better understand the present. Needless to say, it is important to understand the past and present in order to best design the future.

With the advent of the steam engine in 1743, the Industrial Revolution in Europe took on great impetus. However, at that time in the American colonies, you would find an agrarian society engrossed in agriculture, mining, lumbering, and fur trapping. The state of manufacturing was solely in the cottage industry mode and was more than adequate for a primitive society.

Shortly after the Revolutionary War, around 1780, the U.S.A. commenced in earnest its own Industrial Revolution. Emancipation from British rule unleashed a tremendous economic force demanding large scale quantities of goods in the newly-found democratic states. In fact, shortly thereafter in 1792 the Wall Street stock exchange was established.

Later the battle between the North and the South in the mid-Nineteenth Century escalated demand, further proliferating our manufacturing base. This was followed by continuous expansion, while management and labor were developing into two distinct camps. This lead to the formation of the first labor union in 1884 by Samuel Gompers, a New York City cigar maker.

As labor sought recognition, we saw increased management resistance, fostering the findings of F. W. Taylor in 1904. His notion of Scientific Management established a new approach in that field while creating further public recognition for the labor-management separation. There were those who said that Scientific Management dehumanized the work place. Managers could now assume their "divine right" and scientifically specify the output of people at work.

1918 saw the effects of World War I on our industrial society. The demand for goods to service that conflict created further formalization of the manufacturing institution. Labor strife was still rampant until 1934 when the Wagner Act was ratified, giving the worker the right to organize. This was the first document of labor legislation in the U.S.A.

Then followed the war to stop all wars, World War II, in 1941. This conflict created an unprecedented demand for goods. We responded with a level of productivity never experienced before in recorded history. Consequently, we won the war. Then came a reconstruction period with major offshore factories of the world completely devastated.

The 1950s were called the Automation Age. We were in a burgeoning growth mode. Automation had a different meaning during that era. The demands of that economy clearly sought out and put into effect automatic machinery to reduce cost and increase output. Quality was not an issue; it was a given. Top quality was ingrained in every man and woman at work.

Therefore, automation of the 1950s meant installing automatic machinery. Since we were in a growth mode, very few workers lost their jobs. The machinery eliminated labor-intensive work, and the displaced people were transferred to another department that was struggling to keep up with schedules.

It was far less complicated to be an automation engineer in those days. In the first place, "leapfrogging" was an unknown. The pace of developing technology was not rapid. Consequently, competitors were not obsoleting or "leapfrogging" each other's equipment by rapid advancements. Hence, the purchase of a new machine was not looked at as risky.

Second, electronics were not in vogue. Relays, timing devices, and special purpose switches were the leading edge. An implementor engineer could keep pace with technology with very little effort and, consequently, the Automation Age of the 1950s took quantum steps towards making the U.S.A. a global competitor.

The decade of the 1960s was called the Computer Age. Business systems were changing to electronic data processing. Management Information Centers were springing up across the country. At that time, it wasn't even dreamed that computers and manufacturing could conceivably be linked together. The personal computer which was the missing link was yet to make its impact.

The 1970s somewhat diluted the focus on manufacturing with the advent of the Space Age. Astronauts were replacing Wild West cowboys as heroes. Moon walks and the invasion of Mars were all topics of cocktail conversations. Manufacturing seemed sound as ever and destined to safely continue as world class forever. So when the backslide commenced, it went unnoticed.

The 1980s are the era of awakening. We have changed from an

industrial society to an information society. Consequently, we find ourselves in a new Automation Age that requires considerable definition.

There are those who label this era the Age of Anxiety. Drugs and alcohol are a threat to our security. We turned from a trade surplus to a deficit. Unfriendly takeovers and insider trading are no longer unusual. Restructuring and downsizing has led to management massacres. Moral values have suffered. Employee loyalty is waning. Teleevangelists are not exemplary in their behavior. The same can be said for leading government and political figures.

The world around us is highly technical, using a new language, and espousing such anomalies as machine vision and artificial intelligence. This is beyond the ken of average man. It is no surprise that a strong case can be made that the current Automation Age is in truth the Age of Anxiety.

What will we label the 1990s? Will it be the Age of Decadence or the Age of Deliverance? This book is certainly dedicated to providing a solution leading the nation to the latter. But, first of all, we must eliminate the ignorance about the current Automation Age that only serves to deter us from making all of this technology useful.

The 1980s could also be called the Electronic Age, since the proliferation of this technology has been astounding. The computer has invaded manufacturing. Programmable controllers have replaced massive electrical control panels. It is the decade of the chip, supercomputing, and numerous others. Most important, all of this serves as a backdrop to the advent of AMT.

In fact, technology is developing rapidly. With this rapid pace has also come a special language that only the most up-to-date can understand. This creates a further problem in that the language is full of acronyms, many of which apply to specific aspects of AMT.

Additionally, AMT has served as the catalyst to create an alphabet soup which must be explained. Each technology is labeled with two or more letters, leading to a language that excludes outsiders such as nontechnical types. For example, artificial intelligence is AI, and statistical process control is SPC. It is now appropriate to list a number of the new technologies, their acronyms, and a simple definition of each. These definitions are purposely designed in an uncomplex manner. Remember the leader doesn't have to be an expert in AMT, but he must be able to manage the expert.

MATERIAL RESOURCE PLANNING (MRP)

A computerized system for tracking and ordering materials throughout the factory. This program can also be expanded to interface with suppliers. Enhancements in waste control, incoming raw material inspection, hand held terminals at the receiving dock, and many others, are all available in MRP systems.

STATISTICAL PROCESS CONTROL (SPC)

A quality control inspection procedure that is linked to computers. Electronic inspection gauges can be tied to the computer feeding information immediately. All manual calculations can be eliminated, while real time quality assurance decision making becomes a residual of this sophisticated system. It is even possible in some cases to "close the loop" and use this process information to adjust the production machinery automatically.

ARTIFICIAL INTELLIGENCE (AI)

A computerized system that is based on rules of thumb called heuristics. It is used in process tuning or troubleshooting, where production workers make these decisions by experientially learned rules. People who are about to retire can be interviewed by specially trained knowledge engineers. The valuable information they retrieve from the domain expert, the retiree, can be organized into an AI system for future employees.

JUST IN TIME (JIT)

A computerized inventory control and purchasing program that reduce inventories to as little as a fraction of a day. The supplier serves as a warehouser, and his reliability in product and delivery is critical. Large savings can be made in reduction of inventory and work in process.

TOTAL QUALITY (TQ)

A broad concept that not only looks at product quality, but also includes scrutiny of the quality of everything the organization does.

There are a number of different programs available, each approach being different. Training is an important element of these approaches.

MACHINE VISION (MV)

The use of television cameras to replace the human eye best describes this application. This system is integrated with a computer to evaluate what the camera sees. High speed, 100% inspection is possible through this technology. Cameras are now placed on the wrist of robots to increase their intelligence. For example, before picking up a part, it can determine the orientation of that object.

ROBOT

A mechanical device that represents a human arm. The hand is called a *manipulator*. *Payload* is the relationship of what the robot handles, as compared to its own weight. When you program the robot to achieve a task, it is said that you are "teaching" the robot. Repetitive tasks like welding are ideal for this technology.

AUTOMATED STORAGE AND RETRIEVAL SYSTEM (AS/RS)

These systems are used for work in-process, as well as warehouses. They allow you to automatically locate and retrieve a bin or pallet from its resting place in a storage rack. You can also automatically send material to the storage area to a designated location. Inventory information is also linked to computers.

AUTOMATED GUIDED VEHICLE SYSTEM (AGVS)

This system consists of a motorized trucking device that travels in a programmed path with interim stops. They are used to carry out security roles. In this case, they have cameras attached. They are office mail carriers, and in-plant material movers. In the military, they are under experimentation to replace soldiers' tasks, while the police use them to defuse bombs.

COMPUTER-AIDED PROCESS PLANNING (CAPP)

A computerized planning system for the manufacturing process. It is linked to the engineering design department. As the design parameters are established for the product, all other disciplines, i.e., purchasing, manufacturing, inventory control, etc. are immediately involved. This system greatly improves the response time to the marketplace.

COMPUTER-AIDED DESIGN/COMPUTER-AIDED MANUFACTURING (CAD/CAM)

This is an object-oriented (pictures instead of words) personal computer system. The engineer can do his drafting on this system. Plant layout, machine design, facilities design are the types of tasks that can be accomplished here. All information can be "downloaded" to other interested parties.

COMPUTERIZED NUMERICAL CONTROL (CNC)

Numerical control is when a machine is run by a prepared tape. For example, a milling machine is no longer adjusted manually. The tape takes care of this automatically while being linked to a computer.

AUTOMATIC ASSEMBLY (AA)

This technique takes complex manual assembly systems and accomplishes them automatically. For example, threading a needle, tightening a screw to a specific torque, and positioning intricate parts are all potential targets for this approach.

DESIGN FOR MANUFACTURABILITY (DFM)

This approach unifies the efforts of product and process design. It attempts to eliminate the so-called "throw over the wall" syndrome where research, design, and marketing zero in on a product specification, and at the last minute say "make it" when they "throw it over the wall" to manufacturing.

GROUP TECHNOLOGY

This technique identifies the similarity of parts and processes in manufacturing. These families of parts and processes once recognized can be used to reduce costs and develop effective design rationalization and purchasing. Stock levels may be reduced and tooling costs and setup times can be optimized since minor changes to similar parts can greatly simplify the inventory.

COMPUTER INTEGRATED MANUFACTURING (CIM)

The overall marriage of every facet of manufacturing with computerization. With advances in sensor technology, you can "bug" aspects of manufacturing never before possible. This information can be used to drive other departments in addition to the production system. The extent to which the CIM principle is applied is variable by type of business.

MANUFACTURING AUTOMATION PROTOCOL (MAP)

A continuing problem in the Automation Age is that various systems from many suppliers cannot readily talk to each other! MAP is an attempt to develop a universal protocol that will make networking possible for all systems in manufacturing.

TECHNICAL OFFICE PROTOCOL (TOP)

This is an attempt similar to MAP, but in the office area.

FLEXIBLE MANUFACTURING SYSTEM (FMS)

This system is a production process that can change over in minor time requirements, automatically feeds in the raw material, and can change over types of raw material automatically. You can make as many changeovers as you like without effecting costs materially. This description is the most sophisticated level, but there are many variations to FMS. In addition, it is expected that this system be fully integrated with computers.

So goes the alphabet soup of AMT. The examples given by no means cover the entire spectrum, but they certainly cover the highlights. You can now see that the Factory of the Future requires specialized attention and expertise.

Although this chapter will not make you a seasoned technologist, you are now a step closer to feeling comfortable with the Automation Age. As a result, you possess an insight into the meaning of AMT and how it can implement the true functioning of the Factory of the Future.

THE AMT POSITIONING

No component of the organization is discrete, therefore, any discussion of AMT cannot continue without taking a look at all of the disciplines. So why not think in terms of advanced finance technology, advanced marketing technology, or as it applies to any other part of the organization?

In truth, advanced thinking, as applied to the Information Age, has been going on in a meaningful way for the past two decades. Computer integration of finance and marketing, for example, are at present in sophisticated stages and are continuing to improve. The difference is that in this example it has been a strength to strength evolution. The improvements that came about were overlayed on a base that was seen as already strong.

In the case of AMT, it is seen as a weakness-to-strength evolution. Manufacturing has failed to evolve commensurate with the other disciplines. Therefore, we are now looking at the need to overlay AMT on a base that is considered less than satisfactory.

It is said that we have moved from an industrial society to an information society. Manufacturing is seen as not fulfilling its place in that movement. On the other hand, if you look at finance and marketing, you will find a different and more encouraging situation. Artificial intelligence is being used to make financial decisions while payroll, accounts payable, as well as many other functions, have been automated for some time. Marketing data from order processing through market analysis has been on electronic data processing equally as long. The present-day salesman in the field records his activities of the day on a hand-held terminal, immediately updating the marketing strategists at the home office.

While this evolution has been taking place among the major portion of the firm, manufacturing has been taken for granted by the rest of the organization. Also, within manufacturing, the leaders have been "asleep at the wheel."

THE ORGANIZATIONAL DICHOTOMY

When you examine the organization, in retrospect the answer becomes somewhat obvious. Unfortunately, the members of the power structure do not include consideration for the production system in their day-to-day decision-making. Manufacturing is seen as a "sponge" that has unlimited ability to absorb the residuals and can be reshaped as many times as necessary. It has been a member of the team, but only as a nonvoting participant.

The power structure has forced a predominant business culture that places the factory in a subordinate role organizationally, as well as dynamically. Consequently, while others are reveling in the advancements of the Information Age, manufacturing is last in line and still waiting.

However, why project the blame to everyone else. Is operations without responsibility for their plight? Beyond doubt, the manufacturing organization has allowed this to happen. They have evolved into a reactionary role, and unless handled very carefully, are not prepared to do otherwise. It is illogical to expect that the producer community after decades in a reactive, subordinate role can emerge overnight as a proactive, powerful influence on the organization. This will take special nurturing over time.

Consequently, you may expect to find the majority of the organization at odds with the issue of AMT. In the first place, too few are competently familiar with the subject. Even the implementor engineer for many reasons is dragging his feet. If the technologist is only lukewarm, how can we expect the nontechnologist to charge ahead fearlessly? He too needs special nurturing in order to harness the technological explosion.

THE RUDE AWAKENING

The rude awakening is at hand. In spite of all obstacles, AMT is here and growing in complexity. It probably will continue to grow in size and intricacy as opposed to going away. Due to the complex nature of AMT, unfortunately, it does not naturally attract, but instead seems to repel the observer from becoming easily involved.

In spite of these obstacles, AMT is still publicized as the link to the competitive edge. Therefore, there is no turning back; there is no middle ground. We must embrace AMT for it is our potential salvation.

Why then do we see AMT catching hold at a snail's pace? If AMT is clearly our salvation, then where is the momentum? In spite of all that we say, the kind of public awareness necessary to drive the AMT movement does not exist in adequate proportion. The answer lies in a visible directive from the power structure. However, until the leaders of the organization understand the management of technology, very little will happen.

Therefore, the answer depends on a new style of management which is prepared to deal with our future and is constantly influenced by the impact of AMT. A style of management that easily deals with the answer to the question, "What is AMT?" The manager of the future must possess the talent to bring all disciplines of the organization to this window of opportunity. That is the only way to harness the technological explosion.

This new style of management can be appropriately labeled automation-driven management. It must be dedicated to surmount the obstacles that are inherent in our current business culture. It must have the wherewithal to provide the impetus for the rude awakening while putting AMT to good use.

This special kind of leadership and its clarification is the objective of this book. It is necessary to lay out a strategy that will reinstate America to its Number 1 position once again. The technology and innovation is all around us. It is increasing by leaps and bounds on a daily basis. However, it is necessary for us to take full advantage of this movement. Automation-driven management will put this leading edge information to good use instead of standing by while others on foreign shores "beat us to the punch."

We must overcome complacency, ignorance, and short-term thinking. We must teach the manager how to make the correct choices that are compatible with the Automation Age. Then each company will be capable of mastering its own destiny.

CONCLUSION

AMT is the whole, but is greater than the sum of its parts. It has far-reaching implications in all segments of the organization. We have developed a case that supports this idea while calling for change and improvement. The following basic statements are important to reiterate:

1. The link between WCM and the competitive edge is AMT.

2. The components of the organization are not discreet.

3. The impact of manufacturing on the total business is significant.

4. We must understand the present state in order to best design the future.

5. The leader doesn't have to be an expert in AMT, but he must be able to manage the expert.

6. Those in the power structure do not include adequate consideration for the production system in their day-to-day decisions.

7. The power structure has forced a business culture that places the factory in a subordinate role.

8. Manufacturing has evolved into a reactionary role and, unless handled carefully, is not prepared to do otherwise.

9. The nontechnologist needs special nurturing in order to deal with the Automation Age.

10. AMT is, within itself, growing in size and complexity.

11. We must embrace AMT for it is our potential salvation.

12. We need a new style of automation-driven management geared to deal with AMT.

13. The manager of the future must possess the talent to bring all segments of this organization to the window of opportunity.

These excerpts from previous segments of the chapter serve as a basis for a precept that AMT is the whole but greater than the sum of its parts. In fact, it is appropriate to talk about total automation

(TA), a concept that requires all members of the organization to become immersed in creating this futuristic business culture.

Manufacturing is not an island. Instead, it can serve as the nucleus of this futuristic culture with endless interconnecting links to the total business. Marketing can enjoy the best possible product if they are properly interconnected. Research and development can create developments that are more accurate and finance will obtain better information for managing the enterprise if they are properly integrated. Every element of the organization has a vested interest in helping manufacturing become the factory of the future. The result is a total business maturation that goes way beyond one's most exalted dreams.

The vehicle that makes this possible is automation-driven management under the precept of total automation (TA). AMT is the focal point, the central tendency of activity, while TA is the overriding precept that bonds the organization into the singular purpose of achieving the competitive edge.

TA provides the total organization team the ability to cooperatively and accurately assess their future. They can more suitably identify and design their role and structure in an environment that is comfortable with the idea of AMT. Innovation must come from the human resource. The power structure must motivate the organization to constantly search out new ways to achieve the competitive edge. This social impact on technology is the basis for the development of TA, an internalized attitude that will promote a futuristic business culture.

We must deal with a rapidly changing environment. It waits for no one. Therefore, it is critical to catch up with the leading edge, pass it by, and keep it in that position. TA is the super model that addresses the needs of a dynamic organization through advanced technology. That is why it is so pertinent to have answered the basic question, "What is AMT?"

Chapter **4**

The Second Class Citizen

INTRODUCING JEFF WARREN

Jeff Warren played sports in high school, was elected business manager of his class, made the honor roll, and was considered an all-around person. After listening to his guidance counselor, he decided to study mechanical engineering. He was exceptionally good at math and science.

So, after graduation, he matriculated to a top ten engineering college and commenced his studies in earnest. He found that in order to meet his expenses, he had to work part-time. That eliminated going out for sports or joining a fraternity. There was too little time and money, and engineering was a demanding subject.

There were people in his dormitory in business and liberal arts who seemed to have plenty of time to recreate, but not so for the engineers. At times, Jeff wondered if it was worth the effort.

He made friends easily and on weekends could always find a companion to join him for a movie, a few hours of window shopping, or a bite to eat. Whenever he did this, his mind was not entirely at ease, since there was always school work awaiting him at the dorm.

After four years of intensive study, Jeff graduated in 1959, number twelve in the class standing. He had several interviews and found it difficult to make a decision since he had no previous experience and could not differentiate intelligently between the companies.

It was almost impossible to compare the prospective employers, except for financial statistics. However, how it would really be working in that organization was a complete enigma. The compensation packages were similar for three of the companies, so he eliminated the other offers and thought only about these.

He decided on one of the companies because it was in the northeast, where he had been raised. So, in three weeks, he agreed to report to work as a staff engineer working on manufacturing projects.

On his first day, he spent most of the time with paperwork. On the second day, he met his boss and the other five engineers in the group. Their experience ranged from one to six years, and they all seemed pleasant enough. This group was part of the engineering department, which totalled 62 professionals, draftsmen, and technicians.

He was taken on a plant tour by the group leader and found it entirely overwhelming. There were too many things that he had not learned in school. Jeff wondered if his situation was really any different than most recent graduates.

Then he was told to stay with a member of his group for the balance of the week and see how he handled his assignments. Those three days proved socially pleasant, but the experience yielded very little learning. On the second week, he got his first assignment, and things became very interesting. He worked as part of a team on a broad variety of projects. A new packaging line, a chemical processing system, a material handling system, and the modification of a complex machine were some of the projects.

He started working on minor portions of these projects and, after three years, was a project manager. He was also catcher on the department softball team and found his lot satisfying.

He had been dating Mary for his first year at work; they got married and had their first child with plans for a second in the near future. Married life agreed with Jeff and, in general, things were fine.

The company was growing and the big push was to upgrade the processes, reduce cost, and increase profits. It was then 1962, the U.S. gross national product was improving steadily, and the engineering department was busy keeping up with the demands of the marketplace.

Two years later, Jeff was working on the top projects and had opportunity to occasionally participate in meetings with the chief engineer as well as marketing, finance, and research people.

He really didn't understand the interplay between these executives, but on the other hand, he didn't have to get involved, except to provide details about the project. One thing, he noticed, was that the chief engineer would at times take an opposing position. However, in the end, he usually agreed to whatever the others wanted. It just took more time and money to get the job done and sometimes they didn't even get more time.

It was now 1964 and, at this juncture, Jeff felt with a growing family and after five years experience, he should be thinking about his future. He commenced reading the want ads and visiting personnel agencies. There were plenty of jobs available at that time for his type of work. He found that he could increase his salary by 20% if he wanted to move to another city.

A STEP UP

He talked it over with Mary and, considering that there had been no promotions from his present group in five years, they decided to change jobs. So, in 1965, Jeff accepted a group leader's position with the engineering department of another company.

He felt comfortable with the new assignment. It meant dealing with more than one project, as well as the four men in his group. He got more exposure to the politics of the company and had already learned the best way to deal with it.

Whatever they want, give it to them. Make sure the price covers all contingencies. If the available manpower can't deliver on time, "farm out" parts of the project. There is no way you can alter the business plan of the company. As long as you deliver the projects and the business continues unhindered by engineering, they leave you alone.

Jeff's boss, the section manager, counseled him frequently, saying, "Solve problems, don't create them." When Jeff complained about some of the last-minute requests from marketing, his boss would respond with, "Don't be negative. These guys know what they need in the marketplace, and that's what pays our salaries."

In 1968, Jeff finished night classes and received a Master's in mechanical engineering from a local university. He was pushing for a promotion. It was rumored that his boss was leaving. Lo and behold, it happened, and Jeff became a Section Manager in charge of four groups. It meant that now he would have to deal with the other disciplines frequently, but that would not be a problem. He had plenty of experience.

He became friendly with some of the people in marketing and research. In this way, he was able to understand what was going on a little better. They cued him in on their plans, but nothing changed. They still wanted what they wanted when they wanted it.

Jeff asked the marketing guys to pass on their decisions earlier. They told him it was impossible, since they needed every last-minute field response to finalize their thinking. He asked the researchers the same question. Their answer was, "Don't interfere with creativity."

Jeff decided not to fight City Hall. The only way to get ahead is to be a hero and perform miracles.

PLEASE THE POWER STRUCTURE

It was now 1969, and salary increases and bonuses were improving and performance appraisals were favorable. Jeff had found the acceptable method, i.e., please everybody, but make them pay, since with time and money you can do anything. They wouldn't listen anyway, and if he said "no," they would complain that he was negative.

A year later, Jeff was given the opportunity to be the department manager of the maintenance department. He was told this would be good grooming for eventual consideration for the assistant chief engineer's position. He took the assignment. It sounded like a good opportunity, since he wouldn't have to "hassle" with the big meetings in this job.

Just do what you're told, stay away from politics. What a relief. However, that bubble burst quickly. He found that whatever went on in the meetings ended up in his lap. He not only maintained the machines, buildings, and grounds, but he also had charge of all the tradesmen who would make the changes to the process that were dictated by the other disciplines.

Working overtime and weekends was normal in order to keep up with all the special requests. New products, product enhancements, and process speed-ups were all constant agendas keeping the maintenance department in full gear at all times. You needed a special temperament to work in this environment. Jeff seemed to know how to deal with this situation, and he was respected as a competent leader.

The nature of the job was one crisis after another. There was neither time nor budgeted money to travel to technical meetings and expositions. On the other hand, keeping up with the latest crisis was time consuming and absorbed all Jeff's attention. Someday it would change, and the engineering personnel can devote their thoughts to professional development.

CHANGING TIMES

It was now 1975. Five years flew by. Technology was changing. The company had installed a computerized preventive maintenance program which was working very well. In this system, make or buy and repair or replace decisions were made automatically by the computer. Also, the efficiency of each maintenance person on the job was measured by this system.

Programmable controllers were becoming part of the scene, as were all kinds of new electronic gear for running the factory. Minority and female engineers were becoming part of the technology ranks. None of this, however, changed the nature of the day-to-day demands. The power structure stayed the same, and full speed ahead with last minute changes prevailed as before.

A year later, Jeff was asked to take on the Number 2 position in the department as assistant chief engineer. He had been hoping for this opportunity and accepted immediately. No more weekends and 15-hour days. Let somebody else get covered with grease in the trenches.

He could now wear decent clothes, attend business meetings, and be an executive at last. Seventeen years under direct fire in the trenches had taught him many important lessons. He was sure that all that training would help him contribute to the interplay in mahogany row.

In his new role, he was responsible for the success of the four department managers; professional development of all, budgetary controls, and planning the future. The four department managers were totally inundated with keeping up with all the day-to-day crises. In fact, it was difficult to get them to sit through a staff meeting.

Professional development was impossible. Too many projects, changes, and deadlines. Budgetary controls centered mostly around making sure there was enough to take care of the unexpected. Capital budgets were purposely inflated. Overtime and weekend work were anticipated at historically high levels. Plenty of outside consultants and machine shop time would also be necessary.

Planning for the future in the true sense could not exist, except to allow double or triple layers of contingencies to deal with the "expected unexpected."

ASCENSION TO THE THRONE

Jeff became expert at dealing with "white water" on a daily basis. Everyone in the organization found him to be a very positive, cooperative, and results-oriented executive. In fact, in closed circles, they said, "For an engineer, Jeff has a lot of business savvy." After six years of persuasion and compromise (more compromise than persuasion), he was promoted to chief engineer with all the trappings, including a private office not enclosed in glass, with a secretary not shared with others.

He devoted a good share of his time circulating among the various vice presidents, keeping them from losing sight of reality. The classical case that was argued repeatedly begged the question, "When a new machine starts up, why does it take so long to reach full production?" This just delays the "roll out." Jeff knew the answer. If they didn't wait until the last minute and planned the project properly, this wouldn't be a problem, but why talk to a stone wall.

It was now 1982, and Jeff was reading in the journals about CIM, AI, MV, as well as other AMT. Vendors were stopping by, pushing their products, promising all kinds of good things. It was difficult enough understanding them, let alone finding a way to apply their products to his processes.

In the first place, how could he go about learning and keeping up with all this technology? He was too busy as it is keeping up with the business. Second, how could he explain all of this to the powers to be? They would think he had gone off the deep end. The answer was simply to put it aside for higher priorities.

Three years later (1985), AMT really hadn't taken off. Some had tried and failed. It was still difficult to keep up and understand. The media were talking more and more about global competitiveness. Jeff was really not interested, the company was doing well, and nobody was really pushing the issue. Jeff wondered what he should do.

At one of the usual mahogany row meetings, the subject came up. Someone read an article in *Business Week*. It was agreed that it was time to take a look at the subject, but no assignments were handed out.

A JOB WELL DONE

1988 rolled around quickly and very little had been done about emerging technologies. Jeff was now 49 years old, was chief engineer of the company, and had been apparently well received. He had been everything to everybody and felt he was an important contributor. However, what should he do about AMT? It was going to take money to train everybody. Most likely the rigorous nature of the problem would require full-time, devoted personnel. Money for training and head count additions are dirty words.

There was no real indication from top management as to their interest. Everything seemed in a business-as-usual mode. Eventually, this issue might be mandated, and perhaps it would be better to wait. Otherwise, it would be a totally uphill battle. Jeff talked it over with his boss, the vice president of operations, who was not a technologist.

It was decided that they should start a couple of projects to establish that they were involved. For example, they would take a look at robotics and machine vision, without spending too much money. As the boss said, "Let's get our feet wet a little, but don't overdo it until we get better clarification of the issue." Consequently, AMT was not given high priority or anything more than token status. "Business as usual" became the preference, while a "wait and see" attitude prevailed.

In the meantime, Jeff listened to the media and read everything available on the subject. He learned several facts:

1. The implementation of AMT is moving slowly in the U.S.

2. Technology for manufacturing is developing faster than it is being implemented.

3. Manufacturing is seen as having failed to maintain a competitive edge.

Now he reflects on these issues while calling upon his experiences, as follows. It's no wonder that the implementation is moving slowly. For the past 29 years I have tried many times to get the rest of the organization to listen to us. They don't want to listen, they just want to have everything their way. So now we respond by getting the job done according to their desires. What do they expect? We're not going to stick our necks out on risky, little understood technologies. They made us the way we are—totally reactive.

Technology is developing so fast that we can't keep up with it anyway. There isn't sufficient money in the budget to travel to expositions, seminars, and professional meetings. Worse yet, we don't have the time. We need a special new approach to this whole question. You can't stay abreast under current conditions. Consequently, we really haven't moved forward on this whole area of AMT.

The result is that they blame us. They should blame themselves. The big hue and cry is that manufacturing has failed to maintain its competitive edge. Naturally, it is the engineers who should implement AMT. However, without money, support, and, most important, a serious top-down message that AMT is right for us, nothing will happen on our part. We're tired of being second class citizens.

THE END

REHASH THE HASH

The engineer is without doubt the potential implementor of technology. Therefore, it is important to provide him with the necessary resources and support to meet this challenge. Unfortunately, the end result is that AMT is not in a rapid state of implementation. Where have we failed?

The Jeff Warren story tells about the sad situation at the interface where the implementation of AMT could take place. It provides you with insights about the evolution that has created our current unsatisfactory business culture. The engineering/manufacturing types have evolved into a reactive profile while a proactive stance on AMT is not likely to emerge on its own.

Consequently, it's necessary to develop a methodology to make this happen. There must be a way to manage technology that yields a better result than we have today. So let's take another look at the Jeff Warren case history and better define our failure.

Jeff realized after entering college that the engineering student was typically overwhelmed with work, contrary to other areas of study. During high school, this separation did not take place. However, the idea that engineers are sociologically different commences on campus and continues at the work place. That engineers are not entrepreneurial and do not make exceptional managers is a common bias in the organization. Consequently, they are relegated to reactive tasks. This is quite contrary to the situation within our global competition.

When Jeff entered industry, he worked on "things" without gaining any exposure to the total business. The emphasis was on projects and, for several years, there was little contact with the rest of the organization.

This raises the question, "Are we properly directing attention to the neophyte engineers' management development program that focuses on his integration into the total business? Are we also extending this thrust to more senior members of the group?"

After five years on the job, our hero gains some outside exposure and begins to recognize the nature of the organization's dynamics. "Give them what they want" or "You can't change the business plans" seem to be the prevailing rules. Engineering and/or manufacturing are really a service function to the rest of the organization.

Only the leadership of the company have the power to change this dilemma. Top-down, clear messages are necessary in the environment that equality for all prevails. On the other hand, the operations personnel need to take on a proactive profile in an environment that signals to them that proactivity is an expectation.

As Jeff moves up the organization, his performance is measured by his getting projects done without creating obstacles. The greater his responsibilities, the more obvious this becomes. Sufficient time and money to cover the "expected unexpected" is the prime factor. Obviously, this whole syndrome is wrong in today's thinking. We know that our global competitors do not operate in this manner. It is going to take a serious change in the mentality of the power structure in order to effect this change. If this doesn't happen, then the implementation of AMT will take the next two decades, and by then, it will be too late. We are currently in that trend.

Unfortunately, professional development has been gradually eliminated in the budgetary "squeeze." Conferences, trade shows, and seminars are costly, but are really mandatory. It is necessary for the professional technologist to maintain an updated inventory of knowledge. Time and money are essential ingredients of this problem. However, over the past several decades, this area has been looked at all too often as a place for cost reduction. Consequently, the potential implementor is not up to date with AMT, does not feel comfortable with the subject, and is not prepared to take a proactive stance on the issue.

Beyond question, all these problems can be solved if the leadership of the individual businesses are sincere and predisposed to do so. In later chapters, we will address these specific items and lay out methodologies that can serve as the basis for designing suitable programs.

The Jeff Warren evolution is typical of what has happened in the engineering ranks. It has resulted because the leaders do not understand how to manage the technology expert, a prime factor in accomplishing this transfer to WCM.

In fact, there is an additional deficiency in this area: the so-called "expert" is culturally losing his license to hold that title. Emerging technologies are outstripping him of his qualifications to meet the challenge of the 1990s. It is important now for the leader to understand how to manage the expert, and this in itself will encourage him to handle the issues of technology.

The business strategy commences in the marketplace. The customer at the point of purchase is the underlying source for understanding how to design the business plan. The major question is how to remain competitive. The answer usually lies in two areas, product differentiation and price. Both of these elements are directly involved with manufacturing.

Obviously, our business culture is wrong since we do not include manufacturing and engineering at the outset. Manufacturing has been subordinated while all logic cries out that by this method we are destined to fail. The minimal acceptable situation at least gives equal consideration to all disciplines, including the operations area. Unfortunately, we are still unable to accomplish this with ease.

The story about Jeff Warren explains how this condition has slowly crept in over the last four decades. Without question, the organization dynamics of the day have allowed this to happen. It is basic to say that a power structure has emerged that sets aside manufacturing matters as secondary.

However, product differentiation and price, as said before, are the mainstays of achieving the competitive edge. Even though both of these tactics depend heavily on manufacturing diligence, we continue to ignore this issue. We have allowed an unsatisfactory business culture to develop. It is ingrained and strongly embedded and will be difficult to change or remove. Nevertheless, in spite of all obstacles, we have to change our way of doing business.

CONCLUSION

It is time to rediscover production and concentrate on making "things" as well as money. Manufacturing could be a center of excellence, but not until its inhabitants are prepared to do so while, in addition, being accepted as equals. Management has the responsibility to create the kind of environment that is compatible with the implementation of AMT. More important, there must be a clear mandate from those at the top.

There must be a resurgence of the management development process in operations. It is necessary to upgrade the entire cadre of factory custodians. We had best deal with that situation now, or it will deal with us later on different terms.

We need a breed of engineers who understand the total environment. They should know teamwork, understand the business, possess the people skills, while demonstrating competence in leading edge technologies. It is time to minimize the cultural divide between management and engineering.

Needless to say, all these changes must take place while business is carried on in normal fashion. There must be growth and profits through human endeavor. Consequently, all of the needed improvements to the organization and technology should become part of an evolutionary process over time that incurs the least amount of disturbance to profit making. However, we are overdue in getting started.

This evolution can only take place if the current leadership of each business understands what is required, how it can be accomplished, and feels comfortable with the management of technology. Most important of all and contrary to the past, there can be no second class citizens remaining in the organization, for engineering is strategic to survival.

Chapter 5

The Vendor-User Problem

THE ADVERSITY

Until such time as vendors of automation are enthusiastically received by eager users ready to implement emerging technologies, the Automation Age will continue in its lackluster state.

This problem is all around us, in spite of the many trade shows, expositions, seminars, and conferences promoting the subject. The literature is abundant with relevant material, while universities research the frontiers of technology, as well as explore new curricula to train managers for the future.

In the meantime, vendors continue to develop new automation products at an exponential rate, while potential users struggle with an ever-changing incomprehensible situation. In spite of all the symposia, consortia, and compendia, the implementation of AMT continues at a snail's pace.

It is generally accepted that countries having shown favorable long-term growth in productivity have also been the ones that invest in plants and equipment at a higher rate. According to the Bureau of Labor Statistics, the U.S. has shown a decreasing rate of capital investment for some time, while productivity improvement is also seen to be on the decline. In fact, from 1948 to 1965, the annual average improvement in labor productivity was +3%. However, during the period from 1965 to 1985, it fell to +1.7%.

Productivity is as critical to our long-term economic growth as it is to our global competitiveness. Capital investment in equipment and automation is also seen as a relevant factor to our ability to compete and grow economically. Therefore, it can be said that a direct relationship exists between the decline in productivity and a similar short-fall in the investment and implementation of AMT.

Until a satisfactory interaction takes place between the vendor and user of automation, we can expect little change in this trend. Consequently, it becomes important for the managers of technology to understand the adversarial state between these two parties.

It is the popular belief that the U.S. is still the greatest innovator in the world. If this is true, there is then a counteracting phenomenon underway, since we are slow to use all of this innovation. In fact, our global competitors are taking these findings to foreign shores and putting them to immediate use.

The issue of automation can be compared to a funnel. We keep innovating and pouring the results into the funnel. We add expositions and seminars to the mix. Furthermore, the literature is abundant and products are becoming available at a blinding rate. All of this goes into the funnel, but the flow from this universe is but a trickle. The clogging element in the neck of the funnel is the adversarial state between the buyer and seller.

It is important to emphasize that until users buy and vendors sell at a satisfactory rate, the Automation Age is not truly maturing. Without this fulfillment, we continue to see AMT lacking, and the needed groundswell never takes place.

In order to examine the phenomenon, the author has interviewed a number of users and vendors face-to-face. The result of this experience is the basis for this chapter. The content of these interviews is quite revealing and provides several important insights on how to better manage technology—particularly for the non-technologist.

As you will see, the nature of the problem is sociological and psychological as well as scientific. The chapter points out further that the manager need not be a technological expert, but instead he must be able to manage the expert.

First of all, it was evident after many interchanges with executives at all levels of the organization that both users and vendors are well aware of the problem. Furthermore, neither side seems to have the total answer. Unfortunately, the probability is not great that these two parties will form an alliance to solve the problem in the near future. Consequently, it is pertinent that we share these findings, since we can develop from this material an approach that will overcome this undesirable state.

The vendor is constantly faced with resistance from the customer. He is unable to consummate sales at a satisfactory rate. This, of course, is not surprising, except that the underlying resistance is adamant and powerful far beyond expectation. This is a unique characteristic of this relationship and is part of the unusual sociological nature of the Automation Age.

An analysis of this adversity should divulge the features that will help us learn how to solve the problem properly. Therefore, an examination of the nature of this adversarial state will provide the fundamental knowledge leading to the design of solutions.

So our next task is to outline and organize the symptoms and discuss their cause and effect. In this way, we will develop evidence that the buyer-seller problem goes far beyond the direct involvement of the two parties in question. In fact, the failure at this interface is deep-rooted in the total user organization. The major deficiency is the common lack of understanding of the management of technology (MOT).

In some way, all functions and disciplines are involved in this problem, which has evolved through a long-term business cultural development that is doomed to fail. It is imperative that we now review the details of the interviews of the vendors and users, while taking full advantage of what lies beneath this surface.

THE VENDOR PROFILE

The following comments regarding the vendor profile were compiled from the responses given by the interviewees. Although all the comments made were not negative, we are listing only those attributes that contribute to the adversarial relationship. This is done to help us "zero in" on solutions to the problem. It is interesting to note that the vendors were as critical about fellow vendors as were the users.

1. The vendor all too frequently requires "up front" money to determine if his product applies.

2. He is under a sales quota pressure that causes him to be aggressive and prone to overlook details.

3. He is given to promising everything to everybody about the performance of his product. He is inclined to oversell.

4. He is subject to failure in meeting commitments.

5. He sells products as opposed to solutions and attempts to "force fit" his product to your application.

6. Since the purchasing agent seeks the lowest bidder, the vendor tends to "low ball" the price. Later, however, he tends to "add on" items after the contract is awarded.

7. Guarantees are not clear, while integration with existing equipment in the customers' plant is an avoidance issue.

8. He seeks to bypass the engineer and others who are not decision makers.

9. He doesn't truly understand the user's needs and business. He lacks the "partnership" approach and doesn't help the user plan for automation. There is a gap of purpose. The vendor usually takes control of the project because of superior knowledge.

Typically, automation products such as machine vision have unlimited application. Therefore, the vendor requires up front money in order to determine the applicability of his product to the customer's problem. Often the project specification is not entirely clear, and some experimentation is required beforehand. The user finds it difficult to accept the "up front" investment without any guarantees.

"Low balling" the price with later "add ons" is forced by the outdated purchasing technique of seeking the lowest bidder. It is more important to evaluate the capability of the vendor.

Sales quota pressures create a multitude of sins. The seller promises everything, "force fits" the product with the intent of working out the details later, and for expediency bypasses anyone in the customer ranks who can't make a decision. This does not create a partnership relationship, and results in potential failure to meet commitments. Expediency is the motif, while guarantees are unclear, and the vendor attempts to aggressively take control of the project. Under these conditions, the user sees the project as an "arm's length" issue, while the partnership approach suffers. Once the installation is made, the issue of communications with the other existing systems goes unaddressed.

THE USER PROFILE

1. The user frequently doesn't understand the project requirements, which leads to inadequate job specifications. This results in a moving target, since the specifications are constantly being updated.

2. The implementor/engineer is seen to have low status and lacks managerial skills. He is lacking in knowledge about AMT and therefore sees the entire issue as risky.

3. AMT is constantly changing, which results in a concern about obsolescence. The user lacks in motivation and personal initiative to keep updated with AMT.

4. He does not understand the vendor's technical jargon, nor does he know how to plan an automation program.

5. The automation program is not sincere, but more likely a cosmetic issue. Proper staff or funds have not been allocated, nor is there a well thought out plan.

6. Short-term profit orientation with desire for conventional cost justification is a deterrent.

7. There is no state of urgency, and mutual trust is lacking in regard to the vendor.

8. Upper management does not understand the problem. They are ROI-oriented and have other priorities.

9. MIS is struggling to maintain control. They want to specify automation in manufacturing as they have in business systems.

10. Conventional resistance to new ideas exists, causing a "wait and see" attitude to prevail. The unknowledgeable personnel keep their ignorance covert, and continue to do nothing.

11. There is no ready-made place to learn about AMT.

12. The decision maker doesn't get the full picture and rarely sees the vendor.

13. Bureaucracy is a deterrent, and AMT has not been institutionalized. Leadership on the issue is weak. There is no unification of purpose.

14. Profit making has created a false sense of security. The need to be concerned is not clear.

15. Labor and unions are seen as obstacles to achieving automation. It is easier to maintain a low profile.

After reading this list of comments, you can see why the user rejects the vendor as an adversary. In many ways, he is not prepared to deal with the seller.

One major issue is the user lack of knowledge and understanding of the subject of AMT. This obviously puts the vendor in an intimidator role. This syndrome appears throughout the interviews. The technology is little understood, the project is not clearly defined, job specifications are an enigma, and AMT is constantly obsoleting itself.

The vendor's technical jargon is a "turnoff," while conventional resistance to change continues as an underlying obstacle. Superimposed over all this is the problem that no ready-made AMT continuous learning process exists. The user/engineer is seen as low in status and nonentrepreneurial or managerial, particularly in his attitude toward automation. Consequently, his motivational level is low regarding AMT, and he is strongly lacking in personal initiative.

The usual problems of bureaucratic bloat, infested with internal political struggles, exist in the user environment. The economy is good, so urgency and threatening signs are overlooked.

The decision maker is not available to the vendor and is estranged from direct involvement. This condition is likely a result of total insincerity regarding the subject of AMT. Allocations of staffing and funding have not taken place, while a minimal activity is allowed to provide a cosmetic front. True, devoted leadership to the cause is absent, resulting in a state of noninstitutionalization of the issue.

Short-term profit orientation prevails with a conventional ROI approach to cost justification. Organized labor is seen as an obstacle, although the issue has not been tested.

And so ends the saga of the vendor-user adversity. You can see that the problems are varied in nature, but not unsolvable. Each item needs to be addressed, and a process that considers each situation would serve as a great boon to the acceptance of AMT in the factory.

It is unlikely that these symptoms will go away by themselves. In fact, they will probably become more problematic as more technology is released. The expectation is clear that AMT will grow regardless of acceptance.

That is an unusual feature of the Automation Age. For example, less than 20 out of 200 vendors of machine vision are profitable. In spite of this performance, new manufacturers constantly appear on the scene with a product that will solve the problems of the world. Each vendor seeks to develop his product at the cost of the user while working on an application in his plant.

Eventually, through trial and error, he hopes to arrive at a product that has universal application and a ready profitable market. Until he reaches this state, it is not possible to sell the product with a guarantee of performance.

This is where an integrator enters the scene. This person provides a specialized service to the user who doesn't have the knowledge to decide for himself. The integrator will select the components of the system and will stay on during the debugging period for a price. The problem with this approach is that after the integrator departs, the user must possess the ability to keep the system in operating condition. Therefore, it is imperative that the integrator teach the system to the user before his departure.

The integrator approach is best applied on a user base that is generally knowledgeable and capable. He should become part of the factory team to augment expertise that already exists. In this way, his contribution can be fully appreciated with only minimal risk.

AN ANALYSIS OF THE FINDINGS

An analysis of the vendor profile shows that there are five major issues appearing throughout the interviews:

1. Product selling vs. solution selling

2. Sales quota pressures

3. Overselling

4. Credibility lag

5. Nonpartnership view

The old adage that "the customer is always right" will go a long way toward solving the buyer-seller problem. However, this is valid only when the customer is smart. Unfortunately, this is not the case in the situation at hand.

The vendor can be influenced by the potential buyer if the latter person is prepared to assert himself in a way that suggests a knowledgeable and positive attitude toward AMT. In addition, they need to demonstrate a desire for a mutual objective, i.e., to effect a successful installation in AMT.

To have such an objective salient in the mind of the vendor and to be faced with an evasive user, is a far cry from a mutual objective. Unfortunately, this is a common situation, and the result only serves to aggravate the adversity without fostering the implementation of AMT.

If we take a close look at the five major attributes of the seller, we can demonstrate that a smart user can control and influence the situation in a meaningful way. For example:

1. Product selling vs. solution selling. In the absence of an automation product that applies to all situations, the buyer, if knowledgeable and motivated, can take the lead in defining the problem and then engaging the vendor as to how his product may apply. Most vendors are pleased to teach you about their wares, but it is the responsibility of the implementor to define the application and assist the seller in evaluating the prospective. The team approach is the real key with both parties being intelligent and motivated. The implementor should prepare a clear set of project specifications and present these to the vendor. It may be necessary to enlist the aid of the vendor in this preparation as a partnership is approached. The absence of a job specification is an evasive tactic and encourages the adversarial relationship.

2. Sales quota pressure. This type of pressure added to the em-

phatic resistance of the user causes the seller to behave irrationally, which only serves to feed on the already unfriendly state. It is an endless circle, for the seller faces negative buyers who cause low sales performance for him. The sales manager pressures the salesman to get better results. Consequently, he resorts to drastic tactics. He will bypass his usual customer contact and use others in the organization to seek a decision. He will offer lower prices while postponing some of the elements in the proposal. Unfortunately, this reduces the vendor's credibility and creates even a higher level of repulsion.

Once again, a knowledgeable and motivated buyer can convert this situation into a positive outcome. The seller faces the adamant resistance of his customer, but if a close partnership was formed, centered on a common objective, the above irrational acts would not take place.

3. Overselling. Under this tactic, the vendor promises everything to everybody, leading to unclear guarantees or their total absence. The seller, under this condition, tries to take control of the project to protect his interests. Job specifications, as well as other documentation, are vague and often commitments are not met.

A knowledgeable, motivated user could prevent this from happening by providing leadership and insisting on properly defined job specs and guarantees. No matter what the salesperson says, the final written documents are the critical factors. There is no way to alter these written commitments once they are presented properly except by mutual agreement. You can see that the user has an opportunity to promote the partnership arrangement by these techniques. The vendor and user must work as a team, otherwise chaos prevails. The user must take a leadership position in the buyer-seller relationship. Part of that leadership profile is an aggressive, knowledgeable drive towards the successful implementation of AMT.

4. Credibility lag. Everything that has been said up to this point results in a credibility lag, a major deterrent to a successful interaction, between the vendor and user. If credibility lags, then so does trust, and a sale will fail to occur.

Once this takes place, it is difficult to regain a state of mutual trust. Unfortunately, it could happen that the technology you wish to purchase lies with a vendor who falls under the credibility lag. The solution is to conduct your business with this party meticulously by applying good leadership tactics to convert the situation to the partnership strategy.

Make sure that all the proper steps are taken. Clear job specifications, written guarantees, and agreed upon project objectives are the necessary features to force an understanding and commitment between the two parties.

Credibility lag is a manifestation of an arm's length relationship. The best approach is to prevent this undesirable outcome by starting a partnership in its true sense from the very outset.

THE USER STRATEGY

An analysis of the user profile shows that there are five major issues appearing throughout the interviews:

1. Lack of knowledge

2. Evasion techniques

3. Intimidation role by vendor

4. Cost justification problem

5. Lack of top-down focus

We will address them in turn.

1. Lack of knowledge. Without doubt, the largest single contributing factor to the adversarial relationship is the situation where the user/implementor does not have adequate knowledge about AMT. The problem is that it requires a special effort to keep updated. There is no single source for the information, and AMT is a constant moving target.

The objective to keep knowledgeable requires a special initiative. The implementor has to travel to seminars and expositions, read journals, and participate in societies, while the budget must allow for these expenditures. Most important, he must be prepared to overtly admit that one individual can't know everything. Knowledge is a powerful tool in dealing with the vendor. The implementor needs to plan his continuous learning program in an environment that supports the utilization of AMT.

There is no single source for knowledge about AMT. MOT can only be successful if there are educated implementors as part of the management team. So to accomplish continuous learning, plan to spend time interviewing vendors, linking with universities, subscribing to journals and abstract services, attending training sessions, and participating in societies. The learning must be continuous because the new technologies are emerging on a continuous basis.

Once you stop this process, you are on the way to becoming obsolete. The U.S. is the greatest innovator in the world. The extent of AMT emerging can be incredibly mind boggling, unless you plan to deal with it in an organized manner. It is the responsibility of the management team to create an environment that aids and encourages their implementors to establish a continuous learning program.

2. Evasion techniques. Given the lack of knowledge, the user develops evasion techniques to avoid the vendor. Some of the symptoms are a reluctance to provide job specifications or constantly changing them, complaining about the lack of funds, hiding behind the issue of obsolescence, or failing to come up with a project plan. The result is that nothing happens, and AMT remains a secondary issue.

Resistance to change has always been an obstacle to innovation. The status quo is more comfortable than the upset of change. Consequently, it takes very little to make the implementor resort to evasive tactics. The vendors report considerable frustration over this element. As a result, many of them employ the ''bypass'' technique, where they will avoid their usual contact, and move up the organization to find a decision maker to work with. The decision maker, unfortunately, interprets this as unethical behavior on the part of the vendor, and this aggravates the interactive state between the two parties.

3. Intimidation role by vendor. The vendor has no desire at all to be an intimidator, since it opposes his objective to consummate the sale. But it comes out that way, since there is a tendency for the seller to use his special technical jargon, which immediately turns off the buyer.

We have already reviewed the problem whereby the implementor lacks in knowledge. This leads to an absence of understanding between the two parties, and the customer immediately feels menaced. It goes beyond the issue of understanding, since a system for keeping updated on AMT probably is not available in the user's work environment. More important, he hasn't a plan on how he will solve that problem, so it is an exercise in futility.

The issue of AMT is already seen as intimidating and risky. If the subject is not understood and obsolescence is threatening, you can see that a representative of AMT, i.e., the vendor, is seen as a purveyor of impending danger.

A propensity to feel this way also comes about when the internal organization has not made it clear that AMT is desirable and good for the future of the enterprise. Therefore, the potential user/implementor is already predispositioned by his own organizational dynamics to see the seller and what he stands for as a menace. No wonder the buyer-seller relationship has settled out as adversarial.

The engineer/implementor sees AMT as just another item that will be added to an already fully-loaded schedule. The question as to who should get involved and to what level of expense goes unanswered. It's safer to "wait and see," and any advances by the vendor seem to be out of place, unwanted, and threatening.

4. Cost justification problem. Heretofore, all projects were financially examined on an ROI basis. It is suggested that there are better ways to justify projects. Short-term profit orientation continues to be a stumbling block in the AMT path. The literature suggests looking at material conservation, reduction in inventories, and space savings as cost justification areas. These are not new and have been on the scene for decades. It is possible that we have over concentrated on direct labor as the basis for savings. However, direct labor is shrinking as a portion of prime manufacturing costs. It now runs in the 5–10% area, and overhead and material savings are a more fertile field.

AMT should be looked at as a "defensive" item. You need to take this view to preserve the long-term welfare of the enterprise. If a competitor reduces his market price and begins to take away a share of the market, it is common practice to follow suit and take less profit. This is called a defensive investment, and AMT should be viewed in a similar way.

An AMT budget for staffing, training, and capital investment should be established. Also, cost justification should be the first test of project viability. However, failure at this juncture should not categorize the project as a discard. A careful review may dictate that it falls into the category of an absolute necessity for the future.

Often under conventional labor savings approaches, AMT implementation is a negative. The user constantly rejects the vendor under these rules. Until the company establishes more compatible financial policies, a successful two-party interaction will not take place. Quicker response time to the marketplace, as well as turnaround time for the total process, can be valuable results from AMT usage. They may not translate into immediately recognizable cost savings, but they will help you achieve WCM and certainly place you on the path to the competitive edge and increased market share.

5. Lack of top-down focus. The user and vendor both feel that even when corporate automation programs are established, there is too much talk and too little accomplished. Top-down focus is alluded to, but in practice upper management leaves these matters to others, and the result is confusion and a lack of finite direction. U.S. industry suffers from bureaucratic bloat, internal organization struggles, absent AMT leadership, and insincerity. It has been frequently said that existing AMT programs are cosmetic and a questionable need. Perhaps this is true in an economic environment that is classified as unusually on the upside. All the indicators support that premise, so why make heavy investments in an enigmatic technology that may not be necessary.

Several large companies have invested heavily in emerging technologies and, in truth, there have been a number of failures. All the more reason why top-down focus is desirable. It will require the combined best thinking of the total company leadership to deal with this issue. The front line implementor cannot deal with all the related aspects by himself. They say it is lonesome at the top, but the same condition can exist at the bottom.

We are in a state of a false sense of security. The economy continues to expand, unemployment is low, as is inflation, so why complain about competitiveness. The user having this predisposition certainly is not prone to investing in a little understood, soon to be obsolete technology. In his dealings with the vendor, it is probable that he reflects such an attitude further contributing to their problematic relationship. At some point, AMT must become a strategic issue to be championed by entrepreneurial, risk takers. A visionary user can contribute greatly to the solution of the buyer-seller problem. Consequently, both parties must pay heed to the many problems that we have discussed here.

CONCLUSION

In later chapters, we will outline specific programs to harness technology. In so doing, we will include techniques to solve the vendor-user problem. It is mandatory that the reader understand the adverse dynamics within this two-party relationship in order to successfully manage technology. That is the primary objective of this chapter.

MOT for the nontechnologist is our central theme, and the vendor-user adversity is an important part of that problem. Unfortunately, this unfriendly state is overlooked in attempting to set up a program for the implementation of AMT. After reading this chapter, you can readily see the critical nature of the dilemma. A plan that ignores the buyer-seller problem is doomed for at best only mediocrity. It is not possible to attack the competitive issues of the coming decade effectively without the inclusion of these considerations.

Creativity and innovation coupled with appropriate leadership are the major elements that will take us to the competitive edge. The vendor-user partnership will be sensitive to this approach and is equal in importance to all other elements. Yet it is rarely recognized as part of the planned methodology of MOT. The solution lies primarily with the user, for there are a number of activities that will work, given that aggressive user leadership is applied to the adversarial state. This above all should be addressed as part of the basic design of any program for technology management.

Chapter 6

The Culture Busters

THE INTERLOPER

The previous chapters have described the current business culture which has evolved over the past century. With the exception of the Depression Era of the 'twenties, we have seen a continued growth trend bringing us to our present expansionary state. Even the crash of 1987 has demonstrated our ability to overcome shock, remain resilient, and recover quickly. The U.S. is a strong business giant with more economic power than any other nation in the world.

It is said that the U.S. is the greatest innovator in the world. This is just one of many aspects of our business culture. We have the best of all worlds, including free enterprise, good management, and on and on. Most important, we have been able to modify our culture by various means in a timely fashion as the need arises. Perhaps this has been our greatest strength, a constant vigil which recognizes and responds to evolutionary demands.

Evolution is not a given. In its absence we have status quo. This state of stagnation develops gradually over time. It arrives without notice, infiltrates quietly, and takes hold with tentacles that are difficult to dislodge. Status quo can become part of the culture and as easily as it settles in, it is even more difficult to reject. There are many nations in the world that are at a standstill suffering from disevolution. But not so the U.S. History attests to the fact that status quo has not been a problem.

Evolution occurs whenever catalytic agents are introduced. They can be economic, sociological, political, etc. At first, there is a struggle with this catalyst, since it disturbs the existing cultural balance.

Consequently, the agent is better described as an interloper, and there have been many significant ones over time.

Wars are certainly interlopers in the spectrum of evolution. The advent of organized labor and the Industrial Revolution are both good examples of these catalytic agents. They all seem to have a similar pattern, in that the onset is characterized by resistance and a subsequent struggle. Then follows acceptance and the commencement of cultural change. After this comes a settling period consisting of application and refinement of the evolutionary movement. Finally, there is reform and recycle where the whole process repeats itself.

This example is certainly relevant to the Automation Age of the 1980s. We see AMT as an interloper where we are in the state of resistance and struggle which has created a modest level of acceptance and change. We are only in the preliminary stages of acceptance, and the interloper AMT needs to be reinforced in order to create greater impact. Consequently, we have to bust the current culture.

CULTURE BUSTERS

Let us examine the case history of organized labor to better understand cultural development through evolution. In the late 19th Century, as the impetus of the Industrial Revolution was settling, labor began to pressure management for a better work life. Various groups were formed embodying trades or types of work. There was no central theme, and some of the groups were successful and others were total failures in achieving their objectives.

In fact, little immediate progress was made, except that Samuel Gompers, a New York City cigar maker, in 1884 founded the first formal labor union which developed into the AFL-CIO of today. Then began the real struggle. The word "goon" became commonplace. This was the name given to the men hired by management to wage physical war against picket lines, union meetings, and union organizers.

At this point, the interloper was organized labor. It was seen as unwelcome and unfitting by management, and the struggle began. It took approximately a half century to reach a point where organized labor was legalized. In the meantime, the battle was bloody, and unions unwelcome. So the acceptance period actively started in 1932 when the Wagner Act was passed. This first piece of labor legislation gave the worker the right to organize, and so they did. With acceptance came the subsequent period of change. A change from a stormy era to a matter-of-fact attitude that unions must be dealt with.

There were both good and bad examples of labor management relations. However, the movement prospered and fulfilled a need. Wages and working conditions improved as the final step of the evolution came into being. This final stage consisted of further application and refinement during the 1940s to the 1970s.

Now we see a gradual reversal taking place and a new evolutionary cycle forming. There are those who believe that unionism is obsolete and can be replaced by the new version of participative management. This approach accepts the idea that the production worker should participate in management decisions to varying degrees.

Therefore, the idea of management and labor being two distinct bodies is no longer viable. There is only one group, management. Unionism as a business is slowing down. Global competition in some areas of our industry has contributed greatly to this change. This is true particularly in the Rust Belt, for the Japanese and Koreans have devastated our steel industry.

Major steel companies can no longer cope and so have turned over the business to the work force. Organized labor has become questionable under these conditions, since the purpose and objectives of the group significantly change in this atmosphere. There is clearly a new evolutionary cycle in process in this case history.

The history of organized labor suitably demonstrates the argument that four distinct periods exist in the evolution cycle. We have demonstrated the following steps.

Evolution Cycle

Phase I	Resistance and struggle
Phase II	Acceptance and change
Phase III	Application and refinement
Phase IV	Reform and recycle

The manager of technology must understand the nature of evolution and how cultures are established, for without these insights he will have difficulty in implementing AMT. MOT is directly involved with changing cultures from the factory of yesterday to the factory of the future.

The manager of technology must be a culture buster, a manager who has the understanding and capability of planning and coordinating evolution. He should be aware of the four phases, anticipate them, and enhance their movement. In this way, we can be successful in the introduction of AMT and its timely execution.

AMT, THE INTERLOPER

AMT is the interloper in our business culture and is by no means received with open arms. The automation movement is slow and almost dragging its feet. We have discussed several reasons (in previous chapters) why this lethargy exists. The Automation Age is now in Phase I of its evolution, where resistance and struggle prevail. The era of acceptance and change, which is Phase II, is in its formative stage.

We are a long way from the third phase, which is application and refinement. This period would imply maturation of AMT across the universe. Yes, we are a long way from this state, and the problem is that AMT is still an interloper. Of course, there is some modest level of acceptance and change (Phase II), but we are for the most part in the resistance and struggle era.

If we allow the process to take its course, it could take a half century to pass through the entire cycle. Unfortunately, we do not have that much time, for even two decades is too long. It is critical to reach Phase III, the maturation stage, within five years, otherwise we will lose the battle. Consequently, we must move to the application and refinement stage as quickly as possible.

In order to move at this rate, it is important to concentrate on Phases I and II of the evolution cycle. In Phase I, we deal with resistance and struggle. It is important to understand the nature of the struggle and plan a process that eliminates the obstacles, friction points, and organizational deficiencies.

Once this has been accomplished, you have created an environment that will move quickly to Phase II, acceptance and change, in an enthusiastic fashion. So you can see that the prime issue is to convert AMT, the interloper, who is a trespasser, into a welcome and desirable partner.

THE CURRENT CULTURE

The first step is to take a summary position on the current business culture. First of all, we can quickly conclude that the present state has been successful and will continue for some time to come. In fact, this is one of the problems. There appears to be no need to move quickly to change while those with foresight are warning us that change is necessary to challenge the future.

However, the desirable ground swell fails to appear, and the public awareness needed to make this happen does not exist. So we can conclude that a catalytic agent is needed to serve as a prime mover. In order to best examine how this agent can affect our business culture, we have chosen to take a micro approach. It is more practical to consider the universe on a smaller scale and limit our observations to a single business.

Therefore, as we have said before, it is the responsibility of the individual leadership of each business to effect a change in their particular business. This is a more manageable approach with a much higher probability of success. So take a look at your business, and attempt to describe your business culture, particularly, as it applies to AMT and MOT.

It will help to describe a classical case that can serve as a central theme to a large number of situations. In so doing, we will take a look from the top downward. At the top, we have the power structure who can place great influence on the business culture.

Those in the power structure for the most part are not technologists and those that are have lost their relevancy to this area. This group consists of the elite and has demonstrated in-depth business acumen. The earnings per share are constantly increasing while the stockholders exclaim their exultation over their good fortune.

For decades, this group has taken interest in finance, research, and marketing, while all other functions have continued in a corridor of lessening interest. Finance can show us how to optimize profits; research will achieve the product differentiation; while marketing will strategize the final approach to success. Everything else will follow, and it appears that it has. However, should the need arise, we can always move manufacturing offshore to make up shortfalls in profit.

Then comes the *Business Week* syndrome, where the power

structure follows what is current and pushes those issues within their organization. Hence, we see AMT appear on the horizon as a current issue, and it is immediately accepted at the top as good for the business.

Meetings are held, discussions take place, and all major executives are briefed on the need to move forward with WCM. Then everybody returns to their respective work places and passes the word downward. In some cases, a specialized staff is created to deal with this matter, and the power structure returns to the comfort of their board room to conduct business as usual. The feeling that AMT like all other things in the past will fall into place prevails.

In truth, there is a flutter of activity with superficial fad items to create an aura of conformance. The results are but tokens of compliance, and major programs do not really materialize. AMT under this classical approach is at best a "babysitting" issue. The power structure lets "sleeping babies" lie under the pressure of more urgent matters, particularly, when others should be attending to factory issues. Consequently, a perpetuating force never really materializes.

The power structure can now boast of their futurism and innovation since adequate activity is in place to allow for suitable rhetoric at the next stockholders meeting. Ah yes, there is no greater challenge than the business challenge, the excitement of the monthly P&L, the adulation of the loyal stockholders, or the exhilaration of being always right.

CURRENT CULTURE, PART II

Now let us take a look at the other extremity of the culture—the implementor of AMT. In Chapter 4 we described how the implementor arrived at his present state. Unfortunately, the power structure does not recognize the severity of his plight.

In the first place, there has never been a need to show concern about the manufacturing front. Whatever had to be done "got done." Now that the power structure has demonstrated a desire for the enterprise to move toward WCM, the same expectation prevails— it will be done. The reality that never hits home is that the Automation Age is a different animal and harnessing technology in 1990s and forward requires a completely different formula.

In the past, the achievable objectives were less complex and better understood by the organization between the power structure and the implementor. For example, operations management was better versed in the conventional requirements. Whatever came down the line was a manageable issue, but not so with AMT. Manufacturing management is as superficial in these matters as all other members of the organization.

It is no longer possible to give the command "do" and expect that it will be the same as in the past. The implementor has to be handled completely different. He needs to be embraced as a welcome member of the group, otherwise only a sprinkling of tokens of compliance will emerge, while resistance to change will take on esoteric forms.

It also requires the power structure (PS) to change as well as all others. This is the major overall cultural difference that must take place. In the past, the PS has commanded "do" and they have returned to their quarters and remained the same, i.e., completely indifferent to the factory floor. More important, this has been successful according to conventional indicators.

In the past, we have all believed in the old adage, "you make your money in manufacturing." Now we say "manufacturing is the competitive edge." They actually both say the same thing. The difference is that the path to the objective has changed drastically, and unless this is recognized, we are doomed to failure.

The power structure this time must change along with everyone else. They are without conscious thought that they are as deeply guilty of resistance to change as the rest of the organization. In a recent discussion with the leader of a Japanese manufacturing organization in America, he told me that his boss calls from Japan every week and wants to know what he is doing to competitively beat the U.S. When is the last time you heard your PS inquire about the factory floor?

From a top-down view, the leaders must make it very clear that they expect heavy involvement in an AMT program. They need to be seen and heard where it counts on the factory floor. Staffing, funds, budget allocations for travel and training and follow-up on results are all in order.

This message should permeate every board meeting as well as the engineering staff meeting. The idea that AMT is an interloper must be dispelled. If the message is not clear, then insincerity and cosmetology are suspected and progress is nil. The major catalytic agent in changing the current business culture is the establishment of a socio-cultural link between the power structure and the implementor.

It is not realistic to assume that operations management will singularly take care of this linking problem. When the first budget squeeze takes place, the engineers will be told that travel and seminars are on the forbidden list, and the process of human obsolescence will start over again. After all, a business has to make a profit, and the implementor falls in the category of overhead.

The engineer/implementor needs nurturing. He has been undernourished too long. There are four needs that should be fulfilled:

1. Continuous learning process
2. Organizational stature
3. Equitable rewards
4. Management development

CONTINUOUS LEARNING

Human obsolescence is a disease and extremely contagious. It is particularly pervasive in areas where people work with things such as the engineer/implementor. It is even more severe where the fundamentals of the trade are constantly evolving. If the rate of change is at a reasonable rate like in the 1950s and 1960s, then updating is a casual function. However, with the exponential rate of emerging change of the 1990s, updating does not occur naturally. In fact, it becomes an avoidance issue.

The next problem is time. The daily "firefighting" does not allow the significant quantity of time needed to maintain a current knowledge base. It is necessary to determine the best sources of knowledge, and then leave the factory, travel to these sources for learning, and return to the daily firefighting. Furthermore, it is necessary to offer continuous learning at a lesser level of complexity to the rest of the organization so that the implementor can depend on a supportive and knowledgeable environment.

Needless to say, knowledge is not free. Commensurate budgetary allocations will be needed to allow the implementor to avail himself of learning opportunities. It is generally accepted that manufacturing/engineering types travel the least of all segments of the organization. Consequently, a cultural feature has resulted in that travel is looked at in these circles as a burden.

Travel interferes with hobbies and do-it-yourself projects, as well as personal time. By and large, the engineering culture does not include the personal initiative required to get out of the office and search for knowledge. You can see that this condition has been created by both the organization austerity measures as well as the natural profile of the engineer on the job. This problem will take special nurturing.

ORGANIZATIONAL STATURE

We have already talked about the second class citizen. In AMT terms, the implementor is at the bottom of the ladder, particularly, in relation to the power structure. He is familiar with the people in the factory, but rarely sees the faces of the power structure members. Additionally in the rare case that he receives a communication from that level, it passes through many hands first.

The power structure is a myth to the implementor since it seems so remote and far away. It becomes very difficult to measure its degree of sincerity under these conditions. Consequently, the engineer/implementor does not see himself as a professional, but more of a technician. He is the end of the line of communications, the last one to receive the orders of the day, and he sees himself as the forgotten person.

As we have mentioned before, it is critical to establish a sociocultural link between the power structure and the implementor. This is the only way to create the public awareness necessary to launch the impetus for the implementation of AMT in a timely fashion. Any other method may eventually break through but too late to win the race. We don't have decades, we need significant results within five years.

The socio-cultural link requires that all intermediary segments of the organization demonstrate compatibility with AMT. Unless this occurs, the two extremities of the chain, the power structure and the implementor, will continue in their existing estranged mode. Progress will not take place under this condition. Above all, the stature of the implementor will not improve, and he will continue as an outsider to the inner circle of the enterprise.

EQUITABLE REWARDS

Financial as well as other types of rewards should be associated with the AMT flag. It requires a unique effort and ability to turn this situation around. Therefore, there should be a rainbow at the end of the road. So often because of the many negatives already described, there doesn't appear to be any positives connected with an automation program. This can be discouraging.

The highly visible negative features are the risk of obsolescence, as well as concern for project complexity. The status quo is totally upset and the result is an inordinate amount of upset. Long hours, hard work, and plenty of risk cause one to ponder and hesitate before getting involved. It isn't even clear at this point exactly what all this chaos will do for the business.

Special compensation is the area to examine closely. Bonuses, stock awards, options, and special incentive awards are all areas that could make the difference. Unfortunately, in a great majority of compensation plans, these "goodies" are held out for other levels and are not available to individuals in the trenches.

These circumstances can be viewed as a marketing challenge. The product is AMT and the customer is the implementor. The goal is to achieve acceptance of the product under the condition that the customer will become a long-term user. Furthermore, the implementor is by no means sure that AMT will fulfill his needs.

One of the important fundamentals of good management is that there exist a compatibility between the goals of the enterprise and the goals of the individual. Rewards can help cross this bridge. It now becomes critical to sell this product to the implementor and included in that approach is the consideration of incentives like premiums or rebates both analogous to monetary and not to forget nonmonetary awards properly earned.

MANAGEMENT DEVELOPMENT

"Engineers do not make good managers." This is a belief that has prevailed over several decades. Hence engineers should remain as engineers. However, Japan has proven that belief to be entirely wrong.

In fact, several companies in the U. S. have proven that wrong. Nevertheless, we still predominantly believe this to be true. Why not? It eliminates a whole area of organizational competition.

The U.S. all too often even believes that about technical areas, such as manufacturing. That anybody can manage the factory, as long as you have technologists around to carry out the routine stuff, has been the predominant persuasion. With AMT entering the field, it has become considerably more difficult.

It is time to think seriously about planning the careers of the implementors in a more significant way. Why not develop those engineers that desire to compete for other types of positions? Our global competitors have recognized this opportunity and are seeing this plan materialize at present in an important way.

This does not mean that all engineers will make good managers, no more than all salesmen will have this talent. As part of the upgrading of the second class citizen, we strongly recommend the availability of development opportunities. This means that the bosses have to take an active role in making this happen as part of the regular counseling procedures.

Management development usually requires the incumbent to indicate a desire for a career path leading to a specific managerial position in the organization. After suitable counseling and agreement by the superior and subordinate that the goal is reasonable, it becomes the responsibility of the two parties to lay out a program of training and experience to make this happen. However, there are no guarantees in this system.

Management development can go a long way toward providing motivation and a feeling of self-worth. It is also possible that exceptional managers will be identified by this system that are predispositioned to meet the technological challenges of the future and lead us to the competitive edge.

THE STRATEGIC LINKAGE

Now we are ready to launch a sincere effort of culture busting. We recognize AMT as the interloper and realize that we are in the preliminary stages of Phase I, resistance and struggle. Phase II, acceptance and change, has reached only a miniscule activity level. Let's return to the issue of establishing a socio-cultural linkage between the power structure and the implementor. This is indeed the supreme objective and most important strategic linkage in breaking the culture. Therefore, it is of critical importance to clearly describe this desired result after this linkage is in place.

We will start from the bottom up, the implementor. He will become knowledgeable, motivated, and in control of the vendor relationship. He will feel comfortable in making recommendations since the message will be clear in the organizational environment that support for AMT is abundant from all segments of the enterprise.

Rewards are distinctly available as are opportunities to actively participate in outside professional activities. Keeping updated is facilitated while management development is an evident important offering.

In no way do we expect to create prima donnas. Instead, we want to create an atmosphere of achievement, professionalism, and opportunity which is not unlike all other disciplines. The implementor is not moving with the times, and we can help by eliminating these obstacles.

Now let's move up to the intermediate organization level. They have received the message that AMT is a favored issue and that the implementor is the key to the treasure box. Compatible leadership becomes the pattern, and the major preoccupation is to manage the program successfully, leading to the competitive edge. AMT does not take priority over other activities, but instead is intended to enhance them.

The power structure can truly make all of this happen. First of all, they should require evidence that all the pieces are in place. The most important piece is to clearly send the message that AMT is here to stay and to be utilized to the utmost benefit of the firm. At this point, it is sheer folly to turn away from the scene and expect that it will happen by itself.

The issue will need constant followup and assurance techniques to show what is taking place. In so doing, the socio-cultural linkage between the power structure and implementors begins to materialize. As a next step, there is the establishment of a quarterly meeting of a professional nature to advise all interested parties of what is taking place. The leaders of the company should be mandatory participants. The implementors would obtain firsthand feedback from this approach. There would be no doubt as to the support and expectation levels as a result of this highly visible meeting.

Then it appears that a culture buster program should contain three important elements:

1. Focus

2. Process

3. Accountability

FOCUS

Focus on the issue of AMT must exist at all levels of the organization. However, most important is the top-down aspect. It must be obvious everywhere in a consistent, sincere, and pervasive manner. To send the message and return to coveted chambers while relying on others to complete the cycle is insufficient focus. After all, the objective here is to create attention while placing the subject in the limelight. The anticipated result is to develop a level of concentration that is persistent. Unless this occurs, the techniques of culture busting cannot commence its most preliminary stages of activity.

PROCESS

This process must include as objectives all the elements discussed heretofore: the solution to the user-vendor relationship; elimination of the second class citizen syndrome; a back-to-the-basics review; as well as the four needs of the implementor. Add to this list any other elements that seem important to your business, since all cases are individual. With proper focus and as these pieces fall into place, then the culture change will move through the four evolutionary phases.

Maintain a constant vigil over this process. Be assured that evolution is moving through the phases in a timely, healthy manner. Each phase needs adequate time to settle out in order to assure that progress is rational. However, it is important to watch this movement carefully since we do not have decades. Our goal is to make the transition within five years.

ACCOUNTABILITY

The profit and loss system is thwart with accountability measures. The system of reward and punishment is thwart with accountability measures. So why shouldn't culture busting follow suit? An issue as subtle as culture change needs all the help possible. The basic premise is that progress will be enhanced if periodic reports on progress are programmed into the system.

You can't start the process and expect that it will supervise itself, particularly, if you desire a result within five years. Built-in accountability is a perpetuating necessity in the culture busting process.

CONCLUSION

All the ingredients are presented in this chapter to start your culture busting activities as related to AMT and WCM. The same basic approach will be applicable to any other initiative or segment of your business culture. It is based on the notion that the present be examined very carefully. Once you understand the ramifications of the present system and have identified the interloper, you are then prepared to plan and design the future. In order to effect culture change, you follow up with the fundamentals of culture busting as outlined.

The guidelines are purposely generalized to allow for tailoring to the specific needs of your business. However, if the central theme is followed, you should be prepared to reap the rewards. WCM awaits you through the enthusiastic and comprehensible application of AMT. How to get there is now available to you through the culture busting process.

The remainder of this book will be devoted to refinements in dealing with the automation age, an era in which the potential is truly enormous! You will find that once the organization settles into a forward-thinking mode, that innovation and creativity will be significantly enhanced as well. AMT requires thinking into the future since the subject is futuristic within itself. This frame of reference can serve the enterprise through many other synergistic avenues and raise the total level of success of the firm significantly. Culture busting is synonymous with change and progress and is in total opposition to status quo. This is certainly a powerful asset in dealing with the issue of long-term economic growth and competitiveness. For the final outcome of the culture busting process is to empower the individual to successfully meet the challenge of the future.

Chapter 7

Continuous Learning

THE DYNAMICS

We have already discussed the need for a continuous learning process in the Automation Age. This applies to all levels of the organization in varying degrees. The implementor needs in-depth knowledge, while others can settle for a general exposure sufficient to create a comfort zone.

Therefore, all segments of the organization must gain some knowledge in order to participate in the movement towards WCM. There are no exclusions and the production worker should be included. Knowledge is the singularly most critical factor in successfully dealing with MOT and AMT. Once again, let me point out that it is not necessary to be an expert to manage the expert. However, it is necessary to possess a fundamental understanding of the expertise in order to be successful.

Consequently, it is important to devote a chapter to the continuing learning process. This is primarily a "how to" treatise on the subject, since under its current disorganized state, it has become complex to the point of deterrence. If progress is to be made, we must overcome this knowledge deficiency.

It is the purpose of this chapter to set a course of action that will lead us to a satisfactory resolution to this dilemma. The search for knowledge needs strategic coordination, while being advantageously positioned within the AMT effort. Continuous learning will fulfill a need for all while making long-term WCM a reality.

In order to best understand the present, it seems proper to take a look at the past, and examine the critical differences. Starting with the

1950s and 1960s, continuous learning was effortless and never became an issue. It was omnipresent and uncomplicated for several reasons. In the first place, the pace of emerging technology was at a comfortable rate. There was time between innovations to apply and gain experience with them. Consequently, obsolescence was manageable and did not create a major threat.

At that time, computers were not part of the equation. As a result, the whole area was ignored and never had an impact on the manufacturing process. Computers were thought to be applicable only to business systems and were not suitable for the factory floor. Hence, learning about emerging technology continued in its effortless state.

Electronics entered the scene in a meaningful way in the late 1960s. Up to that point, automation was seen as a few electrical controls in a panel, uncomplicated but serviceable for the times. A vendor could teach you what you needed to know in a few visits. Now in the new Automation Age, we go way beyond electronics. In fact, in the 1970s and 1980s, personal computers, sensors, and instrumentation have blossomed into a new hi-tech age far beyond the natural understanding of most people. The learning process is no longer effortless.

Unfortunately, there has been a tendency to ignore all this and let the world pass by. There is a lack of understanding of the new technology which has come about unknowingly. However, once it surprises you, it is difficult to reverse the situation. This turnaround will not happen naturally, and a special effort is the only solution. All of a sudden, AMT has invaded the environment and needs to be dealt with.

However, in order to cope with the problem intelligently, the first task is to play a lot of catch up. For example, it would be foolhardy to attempt to plan an AMT program without knowledge. In-depth understanding of the various technologies must exist somewhere in the organization in order to prepare a credible plan. We recommend that this in-depth "know how" become part of the implementor's inventory, while assuring that the remaining members of the firm have a basic understanding.

In order to commence the "how to" explanation, it seems prudent to examine the full meaning of knowledge as it applies to AMT. At first glance, it is safe to say that knowledge certainly includes all of the advanced manufacturing technologies. However, it is also important that we have the ability to select those opportunities from the list that can make the greatest contribution to your competitiveness. This strongly indicates that an AMT program in general is dependent on how you go about using and selecting these technologies. It is also a specific, tailored treatment for your individual business.

Consequently, the marketplace, competition, and the nature of your process have a great bearing on how you proceed. Now that we have expanded the subject to include these other facets, it becomes obvious that any discussion on continuous learning should, for practical reasons, be broadened to the general topic of technology forecasting. We must take a broad view of this entire area, while focusing on the end point, which is an intelligent AMT program that considers all aspects of continuous learning, including the marketplace, competition, etc.

TECHNICAL FORECASTING

Technical forecasting simply means that you will look ahead as to what technologies you will utilize in the future. Conventionally, this has been a chore relegated to research and development. It has been their responsibility to design the products and product enhancements for future launches. Technology forecasting (TF) has been primarily a product development tool, and this approach has yielded an abundance of successes.

On the other hand, what about process development? TF is certainly a suitable tool for this area as well. In fact, TF is potentially a most pervasive technique. MOT is truly intrinsically linked to TF and cannot function well without it. In addition, both process development and AMT are really dependent on TF. We can safely conclude that process development, MOT, and AMT strengthen their interrelationships through TF. The next question to ponder is how often should this type of forecasting/planning activity take place.

Under the product development and R&D effort, TF becomes a periodic event. Product launches occur periodically and, therefore, the corresponding R&D development program is geared to that schedule. Consequently, TF has a tendency to be forced into that mode. However, this would not be true in relation to AMT, for TF as part of this activity must be a continuous function. In fact, the basic design of this system must provide this kind of focus. A constant monitoring of emerging technologies and their inclusion in AMT planning is an important requirement forcing a system of continuous checks and balances.

The next issue is to clearly establish the objective of TF as related to AMT. The objective of TF in this context is to continuously learn about the emerging technologies on a timely basis, assess their value, and infuse this knowledge into an ongoing AMT plan. So you can see that technical forecasting and continuous learning are like two peas in a pod.

The elements of TF fall in three major categories:

1. What the competition has

2. What you have

3. What you need

WHAT THE COMPETITION HAS

Competitive analysis of products is easily done by collecting samples in the marketplace. It's not so easy regarding processes, since entry to the competitor's plant is not readily available. Nevertheless, there is a lot of information around, and a little honest ingenuity can pay off. You can easily see that competitive process information can be valuable in planning the path to competitiveness.

This reminds me of a case history where a company hired a group of consultants to conduct a competitive analysis of both product and process. It develops that one of the specialists was found collecting all of the seemingly important discarded papers in a dumpster behind a competitor's R&D center. The specialist consultant maintained his ethics and never divulged his client.

WHAT YOU HAVE

The starting point or what you already have process-wise is the next important feature of TF. Surprising as it may seem, this information is not always available in a transferable manner. Somebody in the organization may know, but most likely in an informal way. Taking an inventory of what technologies you already have placed in use can be important. By this approach, you learn where you have pockets of knowledge that can apply to other departments or divisions.

WHAT YOU NEED

In order to determine what you need, it is necessary to also know what is available. This is where true knowledge research must take place.

At this point, it becomes important to establish a method to continually impart technical knowledge to the organization at appropriate quantities and levels. This system must have the characteristics of being easily updated, corrected, and accessed. For example, the assignment of researching knowledge should be placed in a central focal point in the organization. These personnel should constantly review their sources and enter this information into a database that is accessible to all. The knowledge base should be coded and clustered into packages for various parts of the organization. For example, there should be a knowledge code for finance people, marketing executives, etc. The knowledge researcher will edit the incoming information and design the various segments for the organization. The implementor, of course, will have at his disposal the more comprehensive knowledge package.

There are several sources for the necessary information, as follows:

1. General literature

2. Professional societies

3. Expositions and trade shows

4. Seminars and conferences

5. University programs

6. Vendors

7. Government agencies

8. Abstract services

1. General literature. Technical magazines, books, and trade publications are rich in advising the reader on what technology is be-

coming available, how to contact vendors, and articles on selected subjects. A good place to locate this literature is through your public library. Most likely, a good share of this material is already entering your firm, but is not available to the majority. Have your mail room conduct a survey for you.

2. Professional societies. The dues are usually small and the returns are usually tremendous from professional societies. They present seminars, publications, and provide an opportunity to meet with other professionals. They also have a research department which is available to find the answers to your questions. This source of knowledge is probably the best bargain in this series.

3. Expositions and trade shows. Expositions, etc., have become big business. They are conducted in most major cities in the country. You will find them listed in most professional society magazines. They present an opportunity to visit a sizeable number of vendor booths at a convention center and learn about available technologies. Additionally, they usually have a series of tutorial sessions that are conducted concurrently with the trade show. A day or two at one of these events will provide a complete treasure chest of information. There is usually a small registration fee which covers the exposition only with extra charges for the tutorial portion.

4. Seminars and conferences. These sessions usually require from one to five days at the site and consist of a series of speakers and workshops on various subjects. This is a wonderful way to learn. Take lots of notes, bring the knowledge to your home base, and enter it into your central database. You will usually find the investment worthwhile, and the seminar content at the leading edge.

5. University programs. The university is in need of real life sponsors in their industrial liaison programs. They provide their industrial members with a source of knowledge and also a place to contract special research efforts. These partnerships can develop into a worthwhile team effort. If desirable, the industrial member can place resident executives at the campus to manage their projects and also take on a study program. The university scene is an important connection to the knowledge and research effort.

6. Vendors. Vendors are an excellent source of knowledge as well as willing helpers in your AMT program. Naturally, they are interested in selling their wares; however, if their product is a good invest-

ment, why not consider the opportunity. Vendors will be happy to have you visit their factories, talk with their experts, and perform a host of helpful activities without obligations. They are usually very knowledgeable in their field.

7. Government agencies. The National Research Council, the Academy of Sciences, NASA, and the U.S. Patent Office are just a few of the many agencies maintained by our government to provide you with technical information. The costs involved are minimal, while the potential return to you is sizeable. You should establish a network with many of these sources and take advantage of your tax dollar.

8. Abstract services. Abstract services have been around a long time in the medical profession. They examine a great number of journals and publish a compilation of the essence of each article in abstract form. There are both off-the-shelf software and service organizations that will provide you with the same advantages on the subject of automation and AMT. You should subscribe to one or two of these programs as part of your knowledge base. Once again, the knowledge researchers should examine these inputs and code them to the appropriate grouping, i.e., finance, research, marketing, etc.

This list is by no means all-inclusive, but certainly demonstrates the idea of establishing an ongoing knowledge research effort to enable the organization to feel comfortable with the issue of AMT. Also, you will develop the ability to determine what is needed in your Automation program. So, by answering the three major areas, i.e., what your competition has, what you have and what you need, you have fulfilled the TF effort and are well on the way to establishing a continuous learning system.

THE PLANNING METHODOLOGY

You are now prepared to move to the planning stage which must also fulfill the condition of being dynamic and ongoing. AMT planning is not a once-a-year activity. It must be similar to the learning process, i.e., continuous.

First of all, the plan will be effective only if several conditions are met. Each of these conditions have been covered in previous chapters. It is imperative that the firm understand the dilemma of the second class citizen and proceed with steps to eliminate that malady. Second, we must take heed of all of the suggestions present in Chapter 6. If this is done, the environment has already entered into a dynamic state, ready for change, and willing to embrace AMT. Finally, the user-vendor adversarial relationship has been exposed, and techniques should be in place to rid the organization of this sizeable stumbling block. Of course, the establishment of a continuous learning program will greatly enhance positive activity in all of these areas. As said previously, continuous learning is the single most important catalyst to effecting successful AMT programs. Once again, the planning process can only be accomplished under a backdrop of the three major remedial efforts which have been dealt with in previous chapters. They are worth repeating:

1. Remedy for the second class citizen

2. Culture busting

3. Remedy for user-vendor adversity

As part of the planning process, it is time to talk about the AMT team. This group should be tailored to fit your individual organization needs. One common recommendation is that the Implementor in some manner should be part of that team. He will bring to the group in-depth technical expertise and guidance that is essential for effective results. Let me outline a few workable examples.

It may be prudent to name a corporate vice president for AMT and allow that person to form the team and coordinate these activities. There are a number of examples using this approach among the Fortune 500. A steering committee is another approach. This interdisciplinary group works with the various departments and provides the necessary coordination and impetus. These two approaches can apply at the corporate level as well as the divisional level.

The important element is to establish an infrastructure that will deal with the WCM goal on a permanent basis while answering to a high level in the organization. It takes this kind of commitment to make AMT successful. A portion of the AMT team approach should include a visible implementation group who actually performs the front line assignments.

This implementation group ideally is made up of industrial engineers, electrical engineers, and computer scientists. These personnel manage, install, and debug projects, participate in the planning process, and conduct knowledge research, while establishing the continuous learning process. This implementation team is also an integral part of the steering committee that plans the culture change of the firm and also serves as "shakers and movers."

Prior to commencing the program design, a back-to-the-basics review should be conducted. Take heed of Chapter 2 and be assured that the concerns of this chapter are taken care of. Remember, it is foolhardy to build an AMT program that relies on an inadequate manufacturing base. A review of Chapter 2 is appropriate at this juncture.

THE PLANNING STRATEGY

Now that the continuous learning process is in effect, as well as the three remedial efforts, we can move ahead safely with the planning process. The plan should be structured in several layers of complexity. Perhaps five layers is a good middle-of-the-road example. Layer 1 is the least complex with increasing complexity to Layer 5. All layers should be integrated. For example, Layer 1 should fit into Layer 5. Each layer of increasing complexity is actually an extension of the previous layers.

A good example of this principle is demonstrated by a computer integrated manufacturing (CIM) program. The objective here is to strategically place advanced sensors throughout manufacturing, as well as AMT of all types. Since all of these applications are subject to computer control, then it is ideal for collecting information never before available through these computers. This information is usable by the production management, as well as all other phases of the business. Process tuning information in a sophisticated way becomes available to the production supervisor and his workers so they can adjust the system more accurately. Instantaneous production results become available to the planners and salesmen and on and on. When planning this project, each element in the first layer of the plan must integrate with all other layers of higher complexity.

The five layers of complexity could represent a plan covering a time expanse suitable to your business. Let's assume that Layer 5 represents where you want to be five years out in a complete installed and working CIM program. Then the five layers each represent approximately one year of activity. So, this plan lays out the least complex activities for the first year that also fit into the final program five years ahead.

A computerized machine vision installation in Layer 1 to inspect product provides you with a 100% improved inspection, and also provides information that can be used elsewhere in the organization. Since machine vision (MV) is a computer-controlled process, it is possible to set up a means of communication between this system and computers in other departments. For example, up-to-the-minute production information generated by the MV computer can be automatically transmitted to the production scheduling computer, where it enables timely decision making. The use of this information eventually becomes part of the overall CIM program in Layer 5.

The formation of an FMS (flexible manufacturing system) in Layer 3 which is computer controlled certainly adds another important increment to the overall CIM program. Real time information about the operation of this system becomes available to the supervisor, planning department, as well as others. The layering method allows you to approach AMT in a logical, workable manner so that when the final layer is installed, all previous elements are totally compatible.

MOT OUTLINE

We have discussed several steps leading to the planning process which actually make up the primary activities of a MOT program. It seems proper at this point to lay this out in checklist outline form in a suggested logical sequence, as follows:

I. Form AMT team
   A. Vice president of AMT
   B. AMT steering committee
   C. Implementation team
      1. Industrial engineer
      2. Electrical engineer
      3. Computer scientist

II. Establish continuous learning
   A. Technical forecasting
      1. What competition has
      2. What you have
      3. What you need
         a. General literature
         b. Professional societies
         c. Expositions and trade shows
         d. Seminars and conferences
         e. University programs
         f. Vendors
         g. Government agencies
         h. Abstract services
   B. Knowledge research
III. Remedial efforts
   A. Second class citizen
   B. Culture busting
   C. User-vendor adversity
IV. Back-to-basics review
V. Planning process
   A. Layering techniques

IMPLEMENTATION AND TIMING

The final step is the implementation of the plan in a timely fashion. The MOT outline above provides a general sequence to follow. For example, the total process commences with the establishment of an infrastructure. These personnel and their resources will provide the central stable theme needed to perpetuate the program. They should establish a tailored strategy for the business and coordinate that program into a state of reality.

As part of that infrastructure, the implementation team very early in the game should commence knowledge research and the establishment of the continuous learning process. For without early activity in this area, there are several other efforts that cannot start. So therefore, the implementation team will get involved in technical forecasting from the onset. Knowledge equates to enlightenment.

Under a backdrop of continuous learning, the organization is ready to accept the culture busting stage. Chapter 6 deals with this issue in some detail. We certainly need top-down focus at this point, as well as execution of the culture busting process and, finally, an accountability process to assure that progress of the desired type and quantity is actually achieved.

The second remedial step deals with the second class citizen. Here, we must strengthen the implementor with knowledge, and assurance that AMT is real and most important. Also, that a sociocultural linkage between the power structure and the implementor will take place.

The vendor-user relationship at this juncture needs proper attention. Chapter 5 covers this in detail. Remember that until this adversity is eliminated, automation on the factory floor will not be implemented at the desired rate and quantity. While these three remedial efforts are in process, a back-to-the-basics review is in order. Be sure that your existing manufacturing base is sound and can be overlayed by AMT effectively. If crisis management prevails on the factory floor, then it's time to change to an orderly process so that rational thinking can be applied to the AMT program.

At this point, the planning is completed, and the result is a program that centers around the several layers of complexity. You are now ready to install and debug AMT projects on the factory floor in a logical sequence in an environment that is prone to acceptance of futuristic thinking.

At this point, the front line implementation team goes into action. Most likely they are part of the engineering department, while participating in the company AMT team as previously outlined. It is their responsibility to design, install, and debug the various projects. It is also their function to seek project approvals and financial support.

Cost justification for AMT needs a new approach. Defensive money should be set aside to support the WCM goal. The long-term demands a state of competitiveness and immediate savings may not be available. Parochial thinking should not stand in the way of progress. Short-term thinking is a detriment.

On the other hand, there is value to quicker customer response time, shorter turnaround time, less work in process, lower inventories, reduced waste, higher quality, and reduced changeover time. All of these areas should be carefully examined when attempting to seek out cost justification for AMT.

The AMT infrastructure needs to confer with the financial executives of the business and reach compatible understanding on these issues. There are those who suggest that the accounting function is obsolete and should be replaced with a different system. History attests to the worthiness of this profession and manufacturing needs their aid and counsel. Like most things, proper communications between interests will lead to a satisfactory resolve. Obviously, financial executives should be part of the AMT infrastructure.

You can now see how the continuous learning program meshes within the overall AMT program. The basic design of this information system takes into account the different needs of the various disciplines in the organization. The marketing executive through his personal computer can call up his special package of knowledge which would be less technical than the information package for the R&D executive.

This internal communications system should also cover social, economic, organizational, as well as technical, aspects of the Automation Age. Special preparation on the issue of competitiveness and what it means to the business are also included as part of the offering. This approach provides a major contribution to the culture busting process.

Technical forecasting has been discussed as both a product development tool and an aid to process development. This chapter would be incomplete without mentioning how both development processes can continuously learn from each other. The continuous learning process demands that both of these efforts be closely integrated. Both parties should communicate, plan their efforts, and execute their plans as though they are one unified effort. In this way, a cooperative initiative results in quicker response time and greater effectiveness in realizing innovative heights. As mentioned earlier, this has come to be known as DFM or design for manufacturability. Continuous learning thrives on this type of competitive approach throughout the organization. Unfortunately, the prevailing business culture has forgotten how to do this. DFM was standard procedure many years ago, but, unfortunately, we have forgotten too many of these conventional practices of the past. Quality is a perfect example.

When the continuous aspect leaves the learning process, then there is a tendency to allow many good practices to go by the wayside. When learning becomes disorganized and takes on haphazard overtones, then good practices get lost. Our world competitors pay attention to those details and force the learning process into an organized, continuous mode. Herein lies an additional advantage and good reason to pay careful heed to the contents of this chapter. The continuous learning process is the singularly most important element in smoothing the pathway to WCM.

CONCLUSION

This chapter goes far beyond the singular issue of continuous learning. We have included references to the "how to" of the execution of the entire AMT program. Please recognize that each company should build on this instruction, and use their own creativity to tailor the program to fit their individual needs. This material is by no means rigid to the point where modifications are discouraged. However, we do recommend that you consider each and every aspect covered in this treatise.

So take careful inventory of all the features in the MOT outline. They should be given careful consideration in developing your AMT plan. This methodology provides for focused leadership, concentration on technology, and a method of managing in an area that has been heretofore an uncontrolled enigma. This configuration of subjects contains the secret to successful MOT. With the growing realization of competitiveness, this material offers a timely and appropriate prescription for success.

It is important to accelerate this critical mass, since time is running out. We cannot afford to wait for decades. We must achieve competitiveness within a few years. The solution is now available to you; simply follow the prescription we have offered for harnessing technology.

We continue to value the need to prepare this material in a way that suits the nontechnologist. This is a critical part of the thesis and is based on an unwavering personal commitment on the part of the author. After all, the majority fall into the numerically important group of nontechnologists.

The technology battle can be a sweet victory. However, it won't happen unless the total organization is committed and involved. It will in some way change the lives of all the organizational members. If handled properly, this change will be for the better.

Continuous learning expands and strengthens the discovery endeavor. This common system for all members of the firm provides better coordination and leveraging of resources. It parallels all initiatives in the firm. It pervades all aspects of the enterprise while creating an unusual synergism within the organization. It develops technology streams for the future and offers a potential that is clearly enormous. Continuous learning is certainly the single most important common thread throughout the endeavor of technology management.

Chapter **8**

The Strategic Linkage

THE STRATEGIC LINKAGE

We have already made reference in numerous ways to the need for a socio-cultural linkage in the firm between the power structure and the actual implementor of AMT. That is the strategic linkage. In the absence of this connection, it is unlikely that WCM can be achieved.

The implementor is not a single person, no more than the power structure. We are not talking about two people shaking hands and pledging friendship forever. More important is the need to develop a positive tie between the two camps and create a sociological pattern of good will and cooperation. As this situation grows and nurtures over time, we can clearly expect to identify a new business culture—a culture that is compatible with the pursuit of MOT.

It is necessary to establish a receptive environment for this coalition, for in its absence, WCM will progress slowly, if at all. It is, therefore, important to develop this notion in some depth to assure the reader's understanding. Effective technology management is heavily dependent on this comprehension and the ability to successfully achieve this important strategic linkage.

The power structure has been in office for a long time in their present configuration. Obviously, they have been successful for decades and most likely for some time to come. Those that understand the nature of the competitiveness issue believe it is time to change. It is time that the power structure change in the way it interfaces with matters of technology.

Unfortunately, the expectation continues that the old ways can still work. All that is needed is a directive or mandate to the implementor, and it will be done. We have already presented several chapters that explain why this is not the case anymore.

It is also true that neither party, the power structure nor the implementor, are prepared to make this transition. Several programs must first be put into place, and those have been outlined in previous chapters as well.

It is the purpose of this chapter to assure that the reader truly understands the meaning of the strategic linkage, how it works, how to identify it, and how to create it. It is a bond between two different type of forces, each decidedly diverse to each other, but with a common purpose.

The power structure has the major responsibility of the overall well-being of the firm. They understand entrepreneurmanship and how to run a successful business. Part of their ken lies in merchandising, marketing products, and product development. They are also well aware of profit and loss, and financial matters in general. After all, profits are the raison d'etre of business.

In fact they are good businessmen at large. They know that they have to fulfill the customer's needs and must invest, for example, adequate funds in R&D to make sure that this objective is achieved. Even though they are inept at understanding chemistry, physics, and other sciences, there is a common ground with R&D. Once the product is designed, a nonscientific description can be gleaned from the complex data and converted to advertising rhetoric, as well as a sales presentation. Consequently, a bond naturally develops between the power structure and R&D through communications and a compatible end point.

Therefore, marketing, finance, and R&D, as well as several other peripheral service functions, make up the classical power structure. This harmonious end point seems to naturally take place between these functions. The power structure is the major force that steers the ship, and, obviously, the system has been successful.

Now let us take a look at the other camp, i.e., the implementor. This really refers to the entire production system from the receiving dock to the shipping dock. It is the turf where CIM, JIT, MRP, SPC, and other AMT can be planted and includes all personnel from the production worker, supervisor, and engineer, all the way up to the vice president of operations.

A natural bond between the implementor and the power structure does not readily take place. In the first place, the implementors are seen as not truly understanding the marketplace. It is better to tell them what has to be done and let them get on with it. They are not seen as entrepreneurs, and when you are sheltered in a cloistered factory environment, that could easily develop.

Consequently, the implementor's understanding of technical matters does not translate into a common understanding with the power structure. Machines, material handling, factories, etc., are left to the engineers and production managers. They are mundane subjects that serve to produce a product. Of course, it can be expected that these personnel will maintain cost reduction programs and other such efforts, but once again within the cloistered confines of the factory environment. As shown in R&D, manufacturing does not finally translate into a common bond of understanding.

In general, the power structure does not enjoy visiting the production site or listening to presentations on hardware, other than at the curiosity level. For a time the issue of quality also fell under the "arm's length" category, but now with the amount of Japan bashing on the issue, quality is naturally flowing into a strategic linkage. Total quality programs are making a significant contribution here since they purport that quality in management, in general, is a linking factor to success. The power structures are devoting time and money to learning these principles.

The quality assurance function has not been seen as part of the implementor's camp. It is an item that fell in between, but due to its close association with production systems, also fell outside of the power structure camp as well. However, the situation is changing with this emphasis on total quality programs.

How did the separation of the two camps in question, the power structure and the implementor, come about? First of all, it was a natural occurrence for the power structure because of their nontechnical predisposition. Conversely, the same can be said for the implementor side of the equation. The implementor camp allowed this to happen because of their predisposition as well.

If the implementor group is truly more interested in nonentrepreneurial aspects of the business, then perhaps the power structure is correct in their beliefs. On the other hand, why not hold the manufacturing group in esteem for their unique important contribution. Why not establish a common ground and invite them to join the team in the full meaning of the word. In the end, it may become obvious that they are not without some measure of entrepreneurship.

The power structure must show sincerity and a true interest in AMT. They must offer aid and counsel to the implementor leading to the strategic linkage. It is unreasonable to expect that it can happen on any other basis. On the other hand, the implementor has a role to play in this reconciliation. He needs to understand the problem and make his contribution as well. However, the power structure is where this transition should commence for changes in behavior patterns are conventionally established at the top.

Therefore, in order to best demonstrate these principles, we have chosen to offer a series of practical real-life type of case histories on a number of aspects of harnessing technology. The common thread throughout these cases is the need for a bonding material that holds the AMT program together. It is our hope that this approach will assist in bringing together the innumerable techniques covered in previous chapters, while demonstrating the value of that bonding agent, the strategic linkage.

CASE HISTORY A—THE CURRENT AFFAIR

Mel Variegate is the president of the Savory Food Company, producing a large variety of edibles. All of these products are well represented on the shelves of most supermarkets. In spite of small margins in the food business, Savory Foods has become a household institution, and the company has enjoyed consistent profitability and growth.

Mr. Variegate is constantly searching for programs to make his company progressive. He has found that the current affairs in the business magazines are a great source of ideas. At one point, productivity was the issue, followed by total quality programs. Now he continues to see articles on the leading edge, cutting edge, competitiveness, and automation.

He asks his manufacturing head to take a look at the situation and quickly returns to the ivory tower to await the next issue of *Commercial Month,* his favorite magazine. He becomes totally engrossed with the next feature article pertaining to acquisitions, mergers, and takeovers.

By the time this fad wore off, three months had passed, and he called up his manufacturing head to see how automation was coming along. He learned that his engineers were studying some projects regarding MV, AI, and FMS. He asked for an explanation, which turned out to be a failure since Mel Variegate had only a vague idea what the engineers were talking about. He returned to the board room, satisfied that something was going on. He was satisfied that his company was addressing the issue of competitiveness.

At the next business committee meeting, the chief engineer gave a presentation on machine vision and requested a $150,000 budget for testing this technique to see if it applies. The chief financial officer advised Mr. Variegate that this was insanity and could mean the establishment of a wasteful precedent. The president decided to let this one go by and await a more concrete proposal, which never came.

Commentary. Typically, the president assumes that AMT can be instituted like all other issues in the past. There is no sensitivity for the second class citizen syndrome or the vendor-user problem. He also assumes the existing business culture is prepared to accept this futuristic issue. The strategic linkage is absent, and conventional thinking and behavior serve as major deterrents to the achievement of true competitiveness. Consequently, progress is slow.

CASE HISTORY B—THE EXECUTIVE LUNCH

The scene is the executive dining room with all of its trappings. At one of the tables, a conversation is in progress between several executives of various nonmanufacturing disciplines. It goes like this:

(A) Have you noticed the literature on competitiveness?

(B) Yes, it looks like we should be involved.

(C) Yes, but what should we do?

(A) Oh, that's up to the engineers and manufacturing people.

(B) What are they doing?

(A) I'm not sure; it looks like a fairly complex subject.

(C) What's the stock market doing today?

The conversation then trails off to Wall Street as they dig deeper into their double lamb chop.

Commentary. The majority of the organization is without true feelings for the subject. Each party is busy with their own responsibilities and automation is seen as somebody else's job. The lack of understanding for the subject and the absence of a way to learn is truly a deterrent. This example is an accurate incremental description of the current business culture. There is an absence of a state of urgency, and if true MOT as prescribed in this treatise does not take place, the competitive edge may be achieved in approximately two decades. Unfortunately, this is too late. Enjoy your lunch for now; soon somebody else may eat it for you.

CASE HISTORY C—THE INFECTIOUS VACUUM

Mary Tardie is a member of the engineering department in the Savory Food Company. She is aware that the president has asked for involvement in AMT. The chief engineer has assigned her to examine the situation and recommend a logical path. This is her first opportunity to get involved and she is pleased. Her first problem is to know where to start; the field is so vast. She decides to call in three suppliers and see what they have to offer. They come and depart, and the experience is devastating for Mary. She can't understand what they are talking about. That's the last time she will expose herself to that kind of embarrassment.

She reads an article on machine vision and decided this could be a good place to start. The next step was to read a textbook on the subject and then call in a supplier. After touring the plant with him, he recommended an inspection application on one of the production lines, but he would need $150,000 up front to make that determination. Mary submits a proposal offering 100% improved inspection and, consequently, a better product. The proposal is turned down and she doesn't know where to turn.

Mary asks for counsel from her boss. He suggests she attend a seminar and see what's happening. It is now four months since she first received the assignment, and Mary is concerned. She attends a seminar on robotics and decides to follow this up.

Upon her return, she once again calls in a supplier on robotics. He has available just the model she is looking for, i.e., his number 76WX817, general application model, for only $65,000. Buy one and you can test it in a multitude of applications. Mary submits a proposal, and to her surprise, it is approved. Perhaps, robots are not as mysterious as machine vision.

Commentary. The second class citizen syndrome, the vendor-user problem, the lack of culture busting, and a decided lack of the strategic linkage are major problems in this case. Consequently, the implementor flounders. Robots resemble machines which are more familiar to the organization. Also, TV and the movies have dramatized their entry into our environment. Consequently, approvals for these machines make more sense to the businessman-at-large. Recent history has shown that in the onset, robots served as symbols to make the employees realize that the power structure is "heads up" in their thinking. After failure to test successfully, they end up as tax deductions and donations to university laboratories.

There is no clear message of support from the top in this story. Consequently, the strategic linkage never truly materializes. Continuous learning is not established and time continues to pass without an AMT plan being carried out. As far as AMT is concerned, the organization is floundering. This is not atypical by any means.

CASE HISTORY D—COSMETOLOGY

Mel Variegate, the president of Savory Food Company, is preparing for his quarterly state of the union address to his CEO. He is deeply engrossed in preparing his presentation, when he sees on his checklist the item, automation. He remembers that the subject has been assigned to the operations division, and he also remembers approving the purchase of a robot. He ponders as follows. We are growing substantially and profits are exemplary. What else can they ask for? There is no immediate need to make large investments in automation and upset the organization. Who knows how the production workers will respond? Whenever we need to, we can step up the program. Our operations people have always responded well in the past. We must move judiciously and cautiously.

He writes into his presentation a paragraph on automation as follows: We have embarked on an automation program and are now in the exploratory stages. Funds have been allocated for testing of robots at our manufacturing sites. Other applications such as machine vision for automatic inspection are being looked at. It is critical that we keep this program active while strategically applying these emerging technologies. It is important to avoid overtaxing the organization and detracting from our current growth and profitability.

Commentary. Mr. Variegate doesn't really understand MOT. He believes that an AMT program can be accelerated immediately at will. He believes that his organization is prepared to do whatever it wants with this program. His experience tells him that this is the way it should be. There is obviously a lack of interchange between the power structure and the implementor. All the common ills are found in this case, and the players are unaware.

Looking through the eyes of the implementor, he is confused. He has been told that the power structure is interested in automation, but the lack of their presence smacks of insincerity. It is difficult to accurately guess what they really have in mind. A copy of the president's quarterly speech is circulated to the management after the meeting. This serves to support their fear. Automation is on the cosmetic agenda.

CASE HISTORY E—IGNORANCE IS BLISS

The Powerhouse Machine and Tool Company are in the process of automating their operation. Their program is totally informal and is being handled as a regular cost reduction program of the past. Whatever ideas come forward, they are randomly examined by the operations people, justified, and installed. However, visible progress is being made even though the program is as random as to whichever vendor knocks on the door at that particular time.

Most of the operations machinery have computerized numerical control (CNC), where each machine is controlled and run by a programmed tape. Several of the operations have been converted to flexible manufacturing systems (FMS). They have automatic raw material feeds, as well as automatic tool changers, and product transfer devices. The latter removes the partially finished product from the machine and transfers it to the raw material feed system of the next machine. Changeovers involve changing the CNC tape and signaling the computers to do the rest in a matter of minutes.

After completion of this highly profitable installation, they realize their folly. At this point, they wish to go to CIM and take advantage of all the available computerized information on the factory floor. Unfortunately, their haphazard approach did not include the consideration of networking. Since they have a randomized selection of computers, it is now complex to have all of these units communicate with each other. The vendors that provided them with these systems did not mention this possibility. What to do now?

Commentary. True MOT does not exist in the Powerhouse Machine and Tool Company. AMT planning is an unknown. This is a classical description of a haphazard MOT process at an unsophisticated level. Whichever vendor knocks on the door is the next potential AMT project. Lack of in-house knowledge has resulted in a networking nightmare. Is ignorance really bliss?

CASE HISTORY F—THE BALANCED EXECUTIVE

Kirk Jumble is the vice president of operations for the Savory Food Company. He has been with the company for 27 years after completing his economics degree. He started as a supervisor and worked his way up the ladder. He is in charge of the total production system from the receiving dock to the shipping, including engineering and maintenance. He is somewhat confused since the president spoke to him about automation. From general appearances, the boss is not that interested. He brings up the subject once a quarter.

Kirk, after all these years, understands his boss and knows that automation is not a priority item. However, he's positive that something has to be done. So he assigns the chief engineer to do something. In fact, within the next twelve months, he anticipates that they will have two or three projects underway.

An inventory of the situation shows that material resource planning (MRP) was installed over two years ago; MRP is automation and can logically be included. Robot tests are still in process, and statistical process control (SPC) is about to be recommended. So progress is being made.

Kirk returns to his daily chores and makes a mental note to take a look at the situation in two or three months.

Commentary. Kirk obviously doesn't understand the true meaning of MOT. Therefore, he is not in a position to influence his boss. Additionally, he is totally dependent on the whimsical approach to automation by his engineering department. In the absence of a clear understanding of the objective, the planning process appears unnecessary. The power structure has not attempted to create a sincere program; therefore, AMT lies in the nebulous state of the organization. The vice president of operations has difficulty in achieving a full-fledged AMT effort. He is the person with the assignment and must determine the extent to which he should push the project in order to achieve a compatible state with the organization. AMT will not "take off" since the strategic linkage has not been established. Unfortunately, the issue is not understood. In the meantime, Kirk is truly a well balanced executive. For the time being, he will make some peripheral progress without disturbing the equilibrium of the firm. Time continues to run out.

CASE HISTORY G—THE INTERNAL STRUGGLE

Bertha Retento is a research scientist in the Savory Food Laboratories. She has charge of a group working on product enhancements. A typical experiment is to try various levels of each ingredient in a food product and determine the best combination through taste panels.

She has learned through a supplier that all of this laborious experimentation can be eliminated through the use of artificial intelligence. This proposition is very interesting since some of the projects require as many as 3,000 individual experiments on the bench.

Bertha pursues this matter to the point where she presents a proposal to the vice president of R&D, who agrees to air the proposition at the next quarterly business committee meeting. While covering this subject at this meeting, the vice president of management information systems (MIS) queried why his team had not been given the opportunity to quote on the project. He advised that MIS could probably do the job better and cheaper. It seemed wise to hold up the project until this situation was cleared up.

A few weeks after MIS submitted a quote, which turned out to be a less than favorable offering, Bertha decided to award the contract to the outside contractor. At this point, the war of the ages commenced.

The purpose is to absorb overhead, so why go outside? You will get closer attention from insiders. When the outsider leaves you will be without follow-up. MIS can supply continuous consultants. The supplier chosen is not the best in the business. The pressure to use MIS became unbearable. Bertha got fed up and busied herself with other endeavors.

Commentary. The MIS struggle is a classical friction point in the automation effort. The problem in this case is that a clear policy regarding AMT and the position of MIS has not been established. In fact, the problem can be broadened to the condition that clarity on the total MOT issue does not exist. The MIS struggle to maintain control of the AMT effort is an extension of the past. Business systems have been conventionally assigned time-sharing on MIS mainframes. Now, with the PC available, should manufacturing be free to move into AMT independently? The answer lies with each individual business to tailor their own procedures. The main issue is the lack of policy which is another missing increment in the Strategic Linkage.

CASE HISTORY H—THE LOST BATTALION

Jack Forelook is the chief engineer for Savory Foods, answering to the vice president of operations. Jack has a sound engineering background with seventeen years of experience and has been with the present company for nine years. He has been told that his performance as chief engineer is outstanding. He has also been informed that he is not seen as a replacement for the vice president of operations, since his management ability is not broad enough.

Nevertheless, Jack likes to think positively and think ahead. In this view, he has concluded that AMT would be handled better by a three-person AMT team. He makes the recommendation and is given the go-ahead to hire an industrial engineer, an electrical engineer, and a computer scientist to work together as a team answering directly to him.

The team becomes solely project-oriented in an environment that doesn't understand AMT. Progress is slow. Realistically, they settle on an objective of one or two projects completed per year. The team is not highly visible, and the major attention comes from their engineering peers. After two years of obscure success, the members of the team resign, move on to better things, and the effort is considered only a moderate success. The majority of the power structure conclude that there must be a better way.

Commentary. AMT was instituted prematurely. First of all, the AMT infrastructure at a higher level was nonexistent and could not support the front line team effort. The three remedial efforts were not in place. Consequently, the rest of the organization was apathetic. Continuous learning wasn't even discussed, nor was there an agreed-upon plan for AMT.

As a result, the team approached the assignment randomly as a splinter group that lacked a strong home base. Lack of purpose and direction continued to wear away any potential feelings of satisfaction. Consequently, the team was short-lived.

In fact, most of the suggestions contained in the MOT outline were not utilized nor ever became part of their knowledge. The AMT team was piloting in uncharted waters, and it was complete folly to think that such an activity could exist without the proper support. The AMT team in this case was truly a lost battalion.

CASE HISTORY I—RIGHT AS RAIN

Mel Variegate, President of Savory Foods, is now two years older, and it has been that long since he insisted that something be done about automation. The company continues to grow, and profitability is most satisfactory. Savory continues to be a leader in the number of new product introductions each year.

Mel feels the company is moving on all frontiers. The demand in the marketplace is strong, competition is fierce, but Savory continues to maintain a healthy share. However, imports are beginning to make some inroads.

Mr. Variegate sits in his ivory tower and ponders over automation as follows:

Each plant site has one or two completed projects and at least one full-time automation engineer attending to the issue. One of these days, we should organize a plant tour to take a look at the results. It would be a good idea to say something about AMT in the annual report. I must send a memo to the public relations department.

The media continues to warn us about the competitiveness dilemma. We seem to be ready if and when we need to strike. A Japanese automobile manufacturer recently came to town. The local labor market is drying up. We have to keep an eye on that situation.

It's questionable that foreign interests will move over here and enter the processed food business. Oh yes, there is a report in the Food Processors' Magazine that metal containers are now available from Asian manufacturers at a substantially reduced cost. I'll have to set up a task force on that item.

Ah yes, things are looking good. We're obviously doing the right things.

Commentary. All that should be said has been said. However, Confucius says, "To be right occasionally could be accidental. To be right often is commendable. To be right all the time is folly."

CONCLUSION

These case histories have been compiled to help in the comprehension of the several critical concepts fundamental to MOT. You can see that a great portion of the activities facilitating MOT are truly nontechnical. They are primary elements of management, leadership, and organization dynamics. In their absence, the technical portions of WCM will yield only mediocre results.

The strategic linkage is in truth a nontechnical issue. It deals with a receptive environment, supportiveness, and communications, all necessary items in any organization. However, the nature of their application as part of MOT is special. The strategic linkage is an important facilitator to the successful utilization of AMT.

It is sociological and cultural in description. However, in its absence, the technological matters will fail to reach world class proportions. This bond can make the critical difference in the quality of your automation program.

If properly planned and well thought out, AMT can prepare your business to meet the challenges of the future that at present are not even discernable. It is impossible to lay out what will be. However, you can rest assured that if there is a way to improve, somebody, somewhere in the world will take that advantage. Why not join them? The logic is sound and cannot be ignored.

The path is well-defined. It requires commitment and leadership on the part of the leaders of each individual business to apply the proper principles of MOT. After reading this book, you can see that there are several issues that are overlooked in our current U.S. business culture. For the most part, our global competitors understand these matters better than us. This book lays out the formula; it is now available for all.

It is time to convert MOT from a weakness to a strength. For example, unless continuous learning in some form is available, MOT will not gain a position of strength. The methodology for its application has been discussed in detail. The rest is up to you, for knowledge nurtures the strategic linkage which, in turn, creates the proper environment for effective MOT.

World class manufacturing status awaits you. Harnessing technology is your strategic advantage. You now possess the secret to making AMT an internalized collaborative process. Technology can be both a problem as well as an opportunity. You are now prepared to deal with that dichotomy in a meaningful way.

Technology forecasting is synonymous with technology foresight. Foresight starts in the board room; however, it is implemented on the factory floor. It is important to carry out this implementation in a way that underscores innovation.

Competitiveness deals with the whole value chain and the total upstream of business. The importance of this issue is enormous as is its potential contribution. It is both pervasive and profound. The same can be said for the strategic linkage, for it brings the two critical, diverse worlds together so that MOT can be carried out effectively to achieve the competitive edge. There can be a worst case scenario or a best case scenario. The choice is yours.

Chapter **9**

The Test Tube Crisis

THE PROBLEM

As horrible as it may sound, you no doubt have heard it said before—what this country needs is a war. Why would anyone in their right mind make such a statement? Wars are devastating. They are economic disasters and heartrending experiences. Nothing on earth short of defending freedom is worth a war. Yet, they say again and again—what this country needs is a war.

Of course, the speaker does not focus on the war itself, but the postwar effects. Actually a more appropriate statement would be—what we need is a crisis to wake up and unify our nation. This suggests that we are asleep and a case could be made that fits the state of automation in the U.S. We are asleep at the wheel, there is impending danger, and it is time to wake up.

The second part of the statement is that a turning point is needed to unify the nation. A case could be further made that relates to the U.S. and automation. There is an obvious absence of public awareness as well as a groundswell of activity that could carry our nation to WCM status. Yes, we are asleep and segmented regarding the subjects of AMT, WCM, and MOT.

There are actually a number of crucial situations around: the trade deficit, the budget deficit, drugs, the stock market crash, etc. None of these issues have incurred sufficient impact to make the difference even though they threaten the very fabric of our society. One could ask, what is it going to take?

The trade deficit attests to our failure to be competitive and apparently will not disappear quickly. The budget deficit is probably beyond any political party or administration, but for sure we are a debtor nation. The drug situation now threatens our national security. The stock market crash couldn't happen, but did, and taught us that Wall Street is really World Street. All of these situations have happened and are happening, but we still go our merry way. We are waiting for that nonlife threatening crisis to occur that will wake up and unify the nation, while providing the surge of awareness that will carry us to WCM status.

In the meantime, the appropriate crisis does not appear, and a broad spectrum national movement is not in place. However, some activity is in process. For example, a number of companies have invested large sums of money in trying to successfully utilize automation. Another sizeable group have entered into various preliminary stages, while the large majority are either disinterested or at a loss to know what to do. It is the realization of this dilemma that has motivated the writing of this book.

Additionally, it is the nature of this total state of affairs that has determined how *Harnessing Technology* has been positioned. If, in fact, the large majority are at a loss as to what should be done, then there is a need to clarify a methodology that will serve as a guideline to MOT. This book must light up the pathway to automation.

Finally, we need an earth-shattering event that will wake up the populace and catapult them into a state of motivation and action leading to the competitive edge. This crucial event should shake our industrial community into a state of awareness. It would force a new realistic perspective while creating a growing uneasiness. It would force a new realistic perspective while creating a growing uneasiness, the necessary forerunner to a state of urgency.

THE TEST TUBE CRISIS

Well, how long can we afford to wait? Will this pervasive turning point ever come? My advice is, don't depend on it. It's a complete "shot in the dark" and is less predictable than a horse race. However, there is another way, and that is to synthesize your own test tube crisis within your own organization. Design and synthesize a combination of predictable events that will occur if AMT is not achieved. This is your crisis.

After completing this exercise, you may conclude that AMT is unnecessary. This is always possible. However, if the opposite is true, it is time to move quickly to the communications stage and publicize the impending crisis to all members of the organization. It is important that they clearly understand the value of automation to the enterprise, as well as the undesirable consequences of standing by in the waiting line.

They should realize that we are already in a war where missiles and aircraft are unsuitable. In order to be victorious, we must recapture and preserve the competitive edge. Any other approach serves only to perpetuate the "flounderers" who stand by waiting and are at loss as to what should be done.

If we are to win this war, we need fire power, strategy, and a threatening force. The fire power is AMT and the strategy is MOT. The threatening force is the synthesized test tube crisis that will unite and wake up the organization into an elite corps ready to do battle. If any one of these three elements is missing, we are doomed to failure.

The test tube crisis is your own situation in your own company. We must assume that a universal crisis on a national basis will not occur. Anything to date seems to indicate that waiting is a risky strategy. The real crisis of consequence is what will happen to your own business if the proper action is not taken. You can set aside the unification and awakening of the nation, and instead take a critical look at your own situation, then wage a determined war. Establish an atmosphere of urgency and impending disaster where there is no tomorrow and you will become a winner.

The opposite of crisis is complacency. In the absence of a threatening force, there is a tendency to be too content. This can be dangerous in economic warfare, and we are certainly engaged in that kind of conflict. There is danger of a rude awakening when all is lost, and victory is out of reach. Complacency is not a strategy; it is a form of surrender.

AUTOMATION-DRIVEN MANAGEMENT

Once the test tube crisis is established, then the objective of WCM becomes clear. This approach creates "believers" who realistically take appropriate action in the areas we have already discussed. Most important, you have established a sound base under which the strategic linkage can be achieved. Then the socio-cultural environs of your business begins to take on different internal overtones. Projects move faster, decisions are timely, and objectives are achieved in areas that previously were only conversations.

Automation becomes a far-reaching subject and potentially effects change in the way of doing things in every far-flung corner of the organization. Consequently, all attitudes and expectations change to a new set of circumstances. CIM provides an information explosion never before imaginable. FMS enables responses to the marketplace that were only a dream before and DFM enables product designs that increase customer satisfaction, facilitate manufacture, and increase profits. What are we waiting for?

The old order is changing and a new type of management is emerging. Now with the advent of super computing, we can expect expansion of managerial capability that borders on the Herculean. If we continue to stand on the sidelines and reject the current state of automation, just think how obsolete we will be in the not-too-distant future.

The emerging management style will be automation-driven. Hence, it is critical that all organizations learn the proper approach to MOT, get involved in a meaningful way, and be prepared for what is yet to come. We are only on the doorstep looking through the window of opportunity. If we feel that the present is complex, you can rest assured that words will not adequately describe the intricacies of the future.

Automation-driven management is the answer. The nature of the task and tools will demand these managers to be totally immersed in the technology of the day. Superficiality will be obvious and unacceptable in the future. Only those that are prepared will succeed, while the unprepared will lack ability to cope. We are heading for a new set of ideas that will depress the inept and embrace the capable.

Automation-driven management will depend on your involvement now. Super computing will impact the system in a way that will pale the complexion of the introduction of personal computers.

If, in fact, the large majority is standing by, floundering, and at loss as to what to do, we are in real trouble. The emergence of technology waits for no one. That is a problem that we have already discussed. The solution is that we must outstrip the pace of innovation with competence.

Needless to say, these comments most assuredly strengthen the case to synthesize a test tube crisis that approximates the real situation in your organization. This will not only serve to create activity in the area of MOT, but you will be laying the foundation that will enable an appropriate evolution to the institutionalization of automation-driven management.

The current socio-cultural environs of the organization are wrong for the future. We have already made this point, and will continue to do so. The old order must change. It is time to establish the strategic linkage between the power structure and the implementor. The power structure must share their advantages with others to encourage them to implement automation for the future. There is no other way to create an automation-driven management which will lead us victoriously through the economic war.

THE ARCHAEOLOGY OF MANAGEMENT

Archaeology is the study of antiquity. It is the study of human life and culture of the past. This pursuit can logically apply to management. Archaeologists are always in search of a lost treasure, a lost city, or a lost culture. We, too, as managers can learn by examining the lost management cultures of the past.

In fact, it has been said many times that we have forgotten and lost many good management practices of the past. Unfortunately, as we move forward into the future, we become preoccupied with change and, in so doing, lose some of the values along the way.

It is not necessary to change completely. Logic suggests that the best formula is to intermingle the future with the past to develop the best mix.

Nevertheless, there is an argument that can be substantiated that the tendency exists to cast off the old indiscriminately while searching for the new style. As a result, we have forgotten a number of desirable past attributes in management. So, therefore, an archaeological adventure of management in search of lost values could be revealing.

This pursuit is particularly important in examining why we are so slow to embrace the principles of MOT and utilizing AMT. Many of these lost talents can be reemployed to assist in accelerating the harnessing of technology.

Lost Values

Teamwork
Details
Esprit de corps
Imbalanced rewards
Fads vs. tenacity
Strategic thinker oversupply
Political freedom
Best interests of company
Long-term view
Imbalanced organization
Self-criticism
Perspective

TEAMWORK

We are losing and continue to lose our focus on teamwork. Design for Manufacturability (DFM) is a typical example. DFM is now offered as an innovative approach to unifying product and process design. It is true that by using computers, it is possible to network these junctions for automatic communications. You can expand this to include purchasing, the suppliers, etc.

The disappointing element is that we have lost the talent of team working these functions. Decades ago, this was standard procedure even though handled manually. Logic dictated that this was the only suitable way, and it was clearly the accepted procedure. Unfortunately, with time this advantage faded away.

This loss of teamwork concentration is probably related to the changing attitudes of employees in general. Survival by the individual seems the more popular focus. This has been brought about by mergers, unfriendly takeovers, which have led to management massacres. The concept of loyalty to the firm has become diluted. The center of personal interest has become ''me and my safety.'' As a result, group orientation opposes the individual emphasis and teamwork has suffered accordingly.

Why is it not possible for the organizational leaders of all the disciplines to get together and voluntarily support the production function through teamwork and help make WCM a reality. You would think that this typical natural desire would emerge and create a very positive environment for AMT. Unfortunately, it doesn't take place. In fact, we now need excuses to form a team. Quality circles were formed to elicit ideas and suggestions from the rank and file. This should be a voluntary, normal daily interchange instead. Technology groups have been formed under the same premise. Unfortunately, we now need special mechanisms to force teamwork, since it doesn't come naturally. Teamwork is truly a disappearing or lost talent.

DETAILS

It is said that one of the key factors in the Asian success story in manufacturing is their attention to details. Specifications, waste, efficiency, time utilization, cost control, simplification of design, inventories, and on and on. Attention to details is another one of our lost values.

We have forgotten how to do this and instead down load this function to lesser personnel without follow-up on results. The conventional executive profile excludes attending to details as an opposing factor to time management.

If you rummage around the archaeological specimens on this issue, you will find that details were everybody's concern at one time. However, the major chore of upper management was to follow up on an exception basis to let everyone know that "details" are not a forgotten issue. The message came through strongly that accountability was built into the system. Now we have too many lofty strategic thinkers that really do not understand the business. Our global competitors are quite the contrary.

ESPRIT DE CORPS

Esprit de corps is a common spirit of devotion and enthusiasm among members of a group. This feeling comes about after the group realizes that through cooperation, teamwork, and hard work, they can achieve the objective better than anybody else. From these experiences will develop esprit de corps.

This discussion suggests the supreme group orientation while subordinating the individual. This can only work if each individual truly finds compatibility with the objective and derives personal benefit from the achievement.

This whole concept finds less and less application in the current business culture where the individual orientation is emphasized. The idea that there is safety in numbers does not necessarily hold true in the current day environment. Esprit de corps is becoming a frayed concept with questionable application. Esprit de corps and its absence is part of the strategic linkage problem.

IMBALANCED REWARDS

There is a perception among the potential implementors of AMT that their opportunities for rewards are limited. This not only applies to compensation, but other "goodies" like promotion, development, travel, visibility, and education. There is, without question, a general attitude that operations people, particularly, technical personnel, are not good at running a business.

If an archaeology expedition were to start out with the quest to search for the lost values, I believe that the findings would show that this condition has prevailed for some time. However, it appears that over the last few decades, this imbalance has escalated. Once again, failure to alter this situation only serves in a negative way. The implementation of the strategic linkage cannot take place under this deficiency. We have to move the second class citizen from coach to the first class cabin.

FADS VS. TENACITY

The U.S. is a fad-oriented society. Even the cabbage patch doll has worn out its welcome, while only a year ago, people would commit murder or happily maim their adversary in order to procure one. Our industrial society is the same. We jump from cure-all to cure-all. We jump from productivity to quality circles to total quality to MRP to JIT and now SPC.

You might ask why we haven't jumped to AMT. The reason is this topic requires tenacity from the very beginning. Stick-to-itiveness is a lost talent and fad orientation is less demanding.

Our global competitors have the ability to select a technique and work it and rework it and stay with it to make it pay off. All successful applications are for the long-term. The U.S. business culture has lost its ability to do this while a short-term profit orientation prevails.

STRATEGIC THINKER OVERSUPPLY

We have an overabundance of strategic thinkers and too few people that can do the job. In the "olden days" you had to demonstrate that you could do the job first before qualifying to be a strategic thinker. The "new think" says you really don't need to know the entire job, as long as you have people under you that you can manage.

There is certainly some truth in that statement. However, it seems a good idea to understand the basics of the job. A good example is the vice president of operations who studied economics. He demonstrates managerial talent, but doesn't understand technical matters. Unfortunately, manufacturing is a technical matter.

In this situation, AMT will not move expeditiously. The leaders need to become part of the continuous learning process. Yet we train strategic thinkers at the university, and they wish to remain lofty and stay in that position. Details are not for strategic thinkers.

It is the recognition of this situation that served as the impetus to the thesis of this book, i.e., MOT, for the nontechnologist. It is not necessary to be an expert to manage the expert. However, it is necessary to possess a fundamental understanding of the expertise in order to manage the expert.

POLITICAL FREEDOM

The effect of politics in the firm has been a subject of many books. History shows that this aspect of business has always existed. The difference is, however, that in today's environment, it has escalated to unusual proportion and effects most aspects of business in a significant way. In order to survive in the organization today, it is mandatory to recognize and understand the dynamics of politics.

The lack of this capability could affect one's career in a serious way, as well as result in placing undue hardships on the individual. All of the discussions on harnessing technology actually center around dealing with politics in the organization in order to overcome the deterrents that it creates. Adept handling of the political aspects are essential in the successful execution of MOT.

Politics have affected every aspect of the human resource equation since the beginning of time. However, in the current era, the influence of this aspect has reached an impact level that could mean the difference between success and failure in an endeavor. MOT is by no means excluded.

BEST INTEREST OF COMPANY

Decisions are made every day in business. As human tensions rise and impending risk creates insecurity, self-interest becomes confused with company interest. Each decision as it is cast should meet the test of fulfilling what is in the best interest of the company. Unfortunately, this does not always occur.

One could argue that to vacillate over MOT is not in the best interest of the company. Avoidance of AMT and failing to establish a supportive environment for automation also falls in that category. Yet it happens frequently. The primary purpose of this book is to teach how this can be overcome.

So, the next time you are involved in decision making, pause for a moment and ask yourself what the conclusion should be for the best interest of the company. If the answer does not appear to satisfy your personal needs, then it is time to adjust. Not the decision, but yourself.

LONG-TERM VIEW

The quarterly profit and loss statement is the supreme measure of success or failure of a business. Consequently, there is primary focus on this report and the criticism is lodged that we have become a short-term oriented business culture. This view opposes the long-term economic growth motif.

Profits have always been important. The profit motive is indeed the raison d'etre of any business, but it is not wise to ignore the long-term. If you do, an unpleasant surprise could be in store for you. The current literature is abundant with the criticism that our current short-term view is incompatible with the Automation Age.

The key to long-term economic growth is competitiveness and the latter is linked with AMT, MOT, and WCM. If the short-term view prevails then, in fact, we are not prepared to meet the challenge of the future. We are doomed to failure unless we change.

And change we will if we follow the prescription that has been outlined here. All of the necessary steps have been covered with candor and detail. We do not recommend setting aside the profit motive. Instead, we recommend the allocation of personal interest, funds, and whatever other resources are necessary to foster the factory of the future.

In the long run you will preserve the continuance of profits and enable the planning of long-term economic growth. A perfect example of failure to do so is the situation in the DRAM chip industry, i.e., the dynamic random access memory chip. It is the primary memory component of all computer systems. Practically the entire industry has moved overseas due to the inability of the U.S. manufacturer to produce competitively. This has happened in the last few years and the worst is yet to come. The bad part—it could happen to you.

IMBALANCED ORGANIZATION

It has been said that the U.S. business culture is market driven. Consequently, the major emphasis is on that discipline. The marketeers are now the kings of the road. Many years of exposure to the marketplace is believed essential in order to be capable to run a business. So marketeers are selected for the top jobs. As a result, there is an obvious imbalance in the organization.

After marketing, you can take your pick, but always manufacturing and engineering end up on the bottom. Herein lies one of the major problems in achieving the competitive edge. It is doubtful that the implementor will take the lead in the automation role without seeing a change in status.

The power structure will have to rebalance the organization to a more even distribution of power. Furthermore, it will be necessary to redistribute the advantages in a more equitable fashion. The success of automation will depend substantially on the ability to redistribute the share of power, otherwise total fulfillment of automation will not occur.

Our global competitors handle this situation in a completely different manner. All the ills that the unbalanced organization create for us do not exist for them. Engineers are frequently CEOs, members of the board, and leaders in marketing. Perhaps we should learn a lesson from them. The question is, will the U.S. power structure be able to make the necessary adjustment?

SELF-CRITICISM

We have dampened our ability to be self-critical. We no longer take the time to rethink everything that we are doing. The nature of business today does not foster this attribute, nor does it allow for the kind of questioning attitudes of yesteryear. There is a tendency to be more superficial.

At one time, it was common practice to ask how, when, where, who, and what, an excellent mental exercise to find the best way to do the job. It was also recommended that you further question—can it be eliminated, simplified, or combined? All of these techniques have faded into antiquity.

Imagine an attitude of this sort being applied to the question of automation. What would happen to the competitiveness issue under this method of self-criticism? The effects of having lost this talent are deep and serious. However, like most things of the past, we do not appreciate the difference.

PERSPECTIVE

We have lost our perspective and it is no wonder. There are too many conflicting signals in the environment. On one side, we have low unemployment, low inflation, continued economic expansion, readily available credit, and a high standard of living.

To the contrary, we have lost several industries to global competition. The TV industry, semiconductors, steel, approximately one-third of automotive are all examples of this failure. On the basis of this evidence, we should be in mourning.

Hundreds of thousands of jobs have been lost, yet more people are employed today than ever before. On the other hand, there are those who are waiting for the bubble to burst. How long will it be until the bottom falls out?

Our point of view is now confused. Our perspective is somewhat bewildered, and a typical result is the disorder existing in the Automation Age. It is not clear to the majority as to the proper approach and what is best for the long-term advantages.

It is our sincere hope that *Harnessing Technology* will solidify your perspective and light up your pathway to competitiveness. It is critical to evaluate the jumbled signs around us for they seem to bear a note of warning. It is time to once again examine our perspective and develop a decisive plan to subdue the enemy. Economic warfare is not new to us. The problem is we have lost our perspective.

CONCLUSION

Now that the expedition is over, our search for the lost values of management explain a number of reasons for the nature of our current business culture. It is time to intermix the old with the new. For sure, it would be to our favor to bring back some of the old values to help us deal with the current dilemma.

As you synthesize your test tube crisis, it would be worthwhile to include some of the lost talents and achieve a far better mix. Clearly some of the values that have worked in the past will help us assuredly deal with the future.

The need to wake up and unify the organization is paramount. Each business must prepare to meet its turning point and finalize the plan for the future. Of course, in the absence of our ability to deal with MOT, there is a danger that the plan would be inappropriate.

We urge you to carefully consider all that is written here. The likelihood is great that you will find the answer to your enigma within these chapters. It is true that the business environment is becoming less and less comprehensible, and advanced technology is the culprit. Therefore, the nontechnologist is in dire need of help in order to manage the future. This is precisely why we have put together *Harnessing Technology*.

We deal directly with the technique of controlled evolution. It is foolhardy to stand by and let it take its own course haphazardly. Managers are leaders, and it is their responsibility to steer the course of the future. However, in order to do so, you must be properly prepared.

In an age of the technology race, nuclear threat, economic uncertainty, and moral turmoil, the manager needs special preparation for what he is about to face. The knowledge of MOT will fill a large void in that special preparation, since it finally leads you to automation-driven management. This style of management will feel comfortable with the innovations yet to come. The major problem at present is to prepare the executive to cope with the technology already here.

The manager of the future will have to be smarter, tougher, and more aggressive than ever before, but always in a backdrop of a higher level of knowledge. Continuous learning is the tool that makes this possible. The test tube crisis is the mechanism that creates the reaction necessary to force progress.

Time is not in our favor, since we are already tardy. You can look ahead and see a multitude of opportunities awaiting us. You can also look arrears and see many opportunities that have been passed by. This happens because of lack of preparation and involvement.

The test tube crisis is the starting point; the MOT outline should be carried out as the next step. You will automatically become prepared and involved in a way that can only lead to success. This is a clear mandate for management, for there is no other way to truly harness technology.

Chapter 10

Decadence vs. Deliverance

DECADENCE

We are now in the concluding stages of *Harnessing Technology,* a primer on the management of technology for the nontechnologist. The path to deliverance has been clearly described. It is your choice to make, for the alternative is decadence.

It would not be difficult to make a case that the U.S. is already on the path to decadence. All great empires have fallen, so why should we expect to be different? The difference is that we have a choice, and the explanation on how to accomplish that is now available. We must preserve that choice.

Our global competitors do not want to destroy us. That would be contrary to their best interest. The U.S. is an enormous market with tremendous buying power. They know the importance of the preservation of this advantage. Their prime objective is to eliminate our freedom of choice and force us to buy their products. They have already made inroads in that arena.

Furthermore, they want to use our technology before we do. That is not difficult to accomplish. We are already forfeiting this phase of the battle. Our high school students reject science. Our graduate schools are filled with foreign students. The slackening of U.S. patent applications is being made up by outsiders. What will the future hold for this part of the equation?

In addition, we are slow on the pick up of the technology that is already available to us. To the victor go the spoils. We are prematurely giving away the spoils before the victor has been determined.

This is incredible, and certainly not the American way. It must be that the American way is changing.

Our global competitors have reasoned that manufacturing and selling in the U.S. makes sense and increases profits. Consequently, there have been many examples of this strategy, particularly, with the favorable currency exchange. Interestingly enough, Americans who are employed by these foreign investors complain that cultural differences make life unpleasant. The expectations are different on both sides, and never the twain shall meet.

Nevertheless, there is a growing trend where foreign interests are buying major businesses and properties in the U.S. How can this happen? Well, for one reason, these offshore investors are very successful in their own sphere of operation and have the money to pay the price. The dynamics are very simple and need no further demonstration.

The major commodity that the offshore investor can amass is wealth, and money talks in world markets. The pity is that we are giving away the store because we allow our global competitors to take profit at a level far beyond their internal needs. Profits at this magnitude become available for transcontinental expansion, and that is what is happening. Unfortunately, we are creating the opportunity for our global competitors to expand beyond their borders. We simply have not stopped the tide.

Then it becomes a matter of escalation. At some level of ownership of property, businesses, and American-based manufacturing, the global competitor becomes the master. As he strengthens his competitive advantage in his export market and gains more and more of the technology advantage, he can place great influence on our economy.

This is only one step away from extraordinary persuasive powers within our government, and eventually, we will feel this invasion in our everyday personal lives. There is no doubt that this scenario is plausible and possible, for we are about to make it happen. Inadvertently, it could happen as a result of our complacency.

Can you imagine the time when a major portion of our labor market is employed by foreign powers? If the current number of employees of foreign interests are complaining, can you imagine the extent of unhappiness among the populace in the future. It is uncertain whether unionism will have a resurgence in attempting to settle this discontent. Subtle cultural differences are difficult to negotiate. This will set us back to the post World War I era.

Majority taxpayers in our major cities will expect consideration and favor no matter whether they hold an American passport or not. Environmental issues, flood control, and nuclear plant sites will become negotiable items. This will set us back by half a century or more.

They will endow our universities with large sums of money for first refusal on new technology. They will teach our farmers new sciences in agriculture and become worldwide exporters of food and agricultural commodities. They will aggressively pursue alternate sources of petroleum, and disband the petroleum cartel. They will pursue every opportunity that we have failed to exploit, while using our innovation and their tenacious business methods.

The offshore investor will do everything in his power to maintain our economy, but only at restricted levels that optimize his return. He will be very careful to perpetuate our buying power, while staying very much in control of as many areas of U.S. influence as possible. And that is the bad news.

IF WE COULD BUT WE CAN'T

We would win back the consumer electronic industry if we could, but we can't.

There are no greater reality-shattering examples than the "if we could but we can't" type. They are horrible, devastating case histories attesting to the failure of our current business culture. You can't argue with real life examples. Why did this happen? The answer is simply that somebody else can produce a better product at a lower price. This is certainly not an earth-shattering revelation, but what is earth-shattering is our inability to fight back.

It must be that we don't really know how to fight back anymore, because if we could, we would. What is it that we don't know? The answer is potentially all or any part of this book. In fact, you can run down the list in the MOT outline and find the answer. In general, we have not become sufficiently competent in the management of technology to be able to quickly move to the competitive edge.

Why is this true? Why is it that we lack ability in this area? The answer is obviously that we do not understand MOT, nor heretofore have we had available a clear organized way to learn.

Why is it that we cannot move quickly? The answer is that we are too far behind our competitors. We have a lot of catch up to accomplish first. Too many of our managers do not possess adequate knowledge about AMT.

Further, there are a special set of circumstances that surround the issue of MOT which must be understood. For example, the peripheral organization dynamics are special. The nature of the players and their predispositions are unique. The existing level of technology in your business, the same status for your competition, and the extent of change that your organization can absorb without upsetting the equilibrium are all important factors for consideration.

"If we could but we can't" is a lament that, hopefully, disappears from our sphere of business. There are glaring examples sprinkled throughout our industrial community at present. However, hereinafter there is no excuse. *Harnessing Technology* is available to you. If we do not take up the banner and move ahead, the worst will happen. The American dream and our exemplary standard of living will gradually erode to a level below first place. In fact, we will eventually approximate the standards of other countries and fall to mediocracy.

Our global adversaries will not grant us a second chance. We either heed the contents herein, or we will fall further to the rear. The opportunity to forge out in front is singular. Once lost, it will be more difficult to retrieve. That is the nature of economic warfare.

ECONOMIC WAR

In real war, people die for the cause, while in economic war, you die a little over a longer period of time. You don't experience pain, but you lose dignity and have feelings of discomfort. Unfortunately, you don't feel threatened until it is too late. In both cases, you are subject to giving up various kinds of freedom and bear the risk of becoming a prisoner. Obviously, you can't afford to lose in either case.

Confusion, complacency, repression of reality, and the like, all serve to aid the enemy. Needless to say, we are aiding the enemy in a significant way. He could not ask for a less formidable foe. We are surrendering before the major attack.

The primary thrust of this type of conflict is infiltration. Infiltrate the university, marketplace, labor force, innovation stream, or any other facet of the economy. Then you could conclude that economic warfare is an honorable endeavor and that the post war treatment of the vanquished by the victor will be honorable. I would like to propose that there is no such thing as honor in any type or phase of warfare.

So you are now forewarned. The reality is unpleasant, and the potential eventualities are devastating. This alone should provide adequate impetus to rise up and bear arms against the invader. It is time to cease complacency. There is no time to be content, with the enemy already crossing our borders.

Repression of reality is a function of the frailty of humanity. It only serves to put aside "what is" and allows you to deal with the "make believe." The loss of one-third of the manufacturing sector of our GNP is by no means "make believe." It is time to cease the repression of reality.

There is no longer a need to permit any further confusion. In spite of conflicting signals in the environment, there are several obvious serious problems around us, and they are typical of the infiltration tactics of economic warfare. Furthermore, you are no longer lacking in suitable weaponry for defense. *Harnessing Technology* deals with all of these issues and is a complete insatiable armory that will serve you well.

POSTWAR RECONSTRUCTION

In economic warfare, there is no postwar reconstruction period for the loser. This is an unusual characteristic of this type of conflict. In this phase, instead of reconstruction, there is escalation. The loser has to bear a more concentrated onslaught of all the hardships already experienced. The enemy no longer infiltrates, for he becomes more aggressive, more obvious, and more competent. It is at this point that the vanquished suffer subtle losses in dignity and freedom. The freedom to purchase at will is reduced, since the alternatives are reduced. Free enterprise becomes limited in varying degrees, while the standard of living takes on overtones of lesser value. Finally, there is a growing tendency for the caliber of life to diminish, while manifesting reductions in self-worth and dignity.

With time, it can only get worse. Remember there is no honor among economic warriors. With time, you cannot expect to be released from bondage. It can only get worse. With time, you can expect no quarter, only more of the same. Economic warfare escalation has no ending. To the victor go the unending spoils.

If you take sound advice, you will reread this chapter more than once. It outlines the truth; it describes the impending uncanny danger that lies ahead, and leaves you only one obvious alternative. You must seek out the avoidance of decadence at all costs. It is your singular most important mission, and it is critical that you respond to the mandate.

There is no need to delay further. You are the manager, in the right place at the right time, and it is the responsibility of management to ward off the invader. The methodology should no longer be strange to you, and you are equipped to map out a course to deliverance. The avoidance of decadence and the search for deliverance is synonymous with MOT.

ECONOMIC CONTRACTION

According to all indicators, we are in the longest period of economic expansion in postwar history. The truth is that we are in an era of economic contraction and are not even aware. Our economy is like a blooming plant. During a drought, the leaves begin to turn brown. In order to stay alive, the plant must have some water, and an occasional rainstorm satisfies this minimal requirement. However, the browning effect around the edges does not disappear, even though the tree continues to grow. The leaves of our economy are already tainted.

There is a contraction in the types of available jobs. Members of the workforce who have devoted their lifetime to manufacturing endeavors are now being retrained for the service industry. It has been frequently reported that wages in this arena are less attractive, but the worker has little choice. Our overall manufacturing capacity base has contracted. The trade deficit increase is a contraction in our ability to export.

We are no longer free of national debt, and we have slowed down our acceptance of advanced technology. Housing is no longer affordable to the young while they invest their money in expensive automobiles. This substitute investment provides pleasure, but lacks the long-term appreciation.

There are many more signs available, but what we have offered should substantiate the premise. The leaves are "browning off," but growth continues in spite of it. Consequently, economic contraction does not attract our attention adequately. It is typical of the other warnings in this chapter. It is much less traumatic to bask in the favorable areas, while rejecting the true meaning of the unfavorable. The malady is present, and that is undeniable, whether we like it or not. The leaves are "browning off" around the edges in spite of the fact that the plant continues to grow.

Contraction and growth can coexist depending on the nature of the relationship. For example, the economists continue to predict economic growth. However, the symptoms that send signals of contraction are expected to continue. Careful examination of GNP predictions give hint of a continued positive growth, but in a decreasing mode. This could mean that economic contraction is about to blossom out and become obvious. If so, then the infiltrators are strengthening.

Economic contraction can be likened to decadence. All great empires have fallen. Will the U.S. join the list, and are we truly already on the way? Obviously, the attitude of "wait and see" can no longer apply. Instead, the U.S. needs to become aggressive and tenacious, like our world competitors. We must concentrate our total energies on the issue of deliverance.

THE QUAGMIRE OF THE 1990S

We are now in the quagmire of the 1990s. Our future is not clear, and the tools that can significantly help us get out of this predicament are not properly understood. We are in uncharted territory, and the quagmire of the 1990s is best described as a mass of indescribable, unusable randomness. The latter is analogous to AMT when the knowledge to deal with it is not available and understood.

It is possible to wander aimlessly in a quagmire without reaching your destination. It can be dangerous and assuredly uncomfortable. The quicker you get out, the better. In fact, almost anything you can say about this swampy plight is analogous to our predicament in the Automation Age. Yes, we are certainly in the quagmire of the 1990s.

In fact, economic warfare is part of the quagmire. There is no major confrontation, while it creeps up on you. All the negatives are active, while you are preoccupied in the quagmire, and you fail to sense the magnitude of the threatening forces. In the meantime, the negative forces continue to attack the very foundation of our society.

Automation needs to be bonded by a set of varied forces. They are social, economic, cultural, technical, and commercial. Until this bonding takes place in the correct combination and allocation, MOT will continue in its bogged down state in the quagmire.

Economic growth is our long-term expansionary objective. The social aspect deals with the standard of living and the effect on people. Our business culture needs to modify in order to create a supportive environment to the technical forces that must be studied and utilized. Finally, we attain a commercial state, where the enterprise is competitive and successful.

When this combination of forces are mature and effective, then the movement towards WCM status will take place in earnest. Then, and only then, will we find our way out of the quagmire of the 1990s.

THE UNFORTUNATE EXAMPLE

The best overall example of our dilemma is the unfortunate federal budget deficit, which is the most damaging, out of control, economic situation at present. It is without doubt the most pervasive problem in this sphere we have. It affects everyone and everything today and into the future.

The budget deficit, in simple terms, is a race between revenues and spending in our federal government. Unfortunately, spending is the winner. We are spending more money than we are taking in. Ordinary businesses would be bankrupt in this scenario, but our government has a unique credit rating, as we continue to go further and further into debt. In fact, our annual interest payments approximate one-third of the revenues received from our income taxes. The present path is a lose-lose situation.

Our excessively unfavorable trade deficit is directly related to the budget deficit. The latter is a major cause of higher interest rates, which dampen investment in capital goods. Businesses need to expand as part of competitiveness. However, if the cost of money is a deterrent, so goes expansion, which negatively affects jobs, productivity, and our ability to export. The trade deficit attests to this problem.

Higher interest rates reduce the propensity to consume for the citizenry. Education, homes, cars, all become more costly. At this point, the standard of living begins to erode for each and everyone of us, and the federal budget deficit is the culprit.

Eventually, this situation will affect inflation, and the standard of living will continue to erode. At some point in the cycle, we will have to deal with recession, and economic expansion will disappear. All of these factors are interdependent and create an unsavory chain reaction.

In order to solve this problem, there are two major alternatives. They are to increase revenues or decrease spending. On one side of the equation, increased revenues equate to higher taxes with further damaging effects to business and the individual. On the other side of the equation, decreased spending means cutting back on government programs. Since a great number of these programs provide aid to voters, the path to cut back is not clear to our lawmakers.

We continue to wallow in the quagmire, wandering aimlessly in an unfriendly environment, while the solution to competitiveness and true long-term economic growth really lies in the hands of our industrial leaders. This is particularly true now that MOT for the non-technologist is recorded and available. The path to deliverance is no longer out of reach.

DELIVERANCE

Deliverance means rescue and liberation. This entire treatise deals with the step-by-step process with which to achieve liberation. We want to be rescued from being second best and mediocre. If democracy is worthwhile and free enterprise practical, then we want to rescue these aspects above all. Any other approach will slowly dilute and restrict the true meaning of these ideals.

Foreign wealth is already zealously buying up the economy of our promised land. With ownership has come an influential outsider in our midst that is in conflict with our culture and intellectual endeavors. He is able to incur political influence from his internal vantage point far above any previous levels experienced. Washington must listen to the voters for sure, but there now is another group of increasing importance that demands their ear.

Economic power eventually equates to political power. Our global competitors are masters at nurturing this strategy. Unless we quell the tide, the American dream will become unrecognizable. We must plan and execute our deliverance from this disastrous possibility.

Liberation implies rescue from a state of indenture, i.e., being placed in a situation against your will. This probably accurately describes the situation. The loss of competitiveness is occurring against our will. The exportation of technology and the inadvertent contraction of the economy are all happening against our will. We are obviously not in a state of liberation, and deliverance is yet to be achieved.

If we took a popular survey of the average American in the street, we would probably find great disparity of opinion over these issues. If we also took the same kind of survey in Washington among our lawmakers, we would probably also find a wide disparity as well. There is an absence of a unified public awareness as to the severity and true nature of the problem. This observation continues to support the idea that it is time for the individual leaders of our industries to take charge and create the necessary change. They are the rescue team of the future and without them, deliverance will not take place.

There is no doubt that we are guilty of several shortcomings. We cherish our freedoms and assume they will always stay the way they are. We cherish our comforts and take them for granted. We know that we have the greatest country in the world and assume that it's a given. The free lunch has disappeared, as has the five-cent cigar. Our freedoms, comforts, and greatness are not guaranteed, unless we take major steps to preserve them. It is time to take action and achieve deliverance from these shortcomings.

Our society is trending towards a preoccupation with the search for pleasure and the avoidance of pain. The pleasures of life are in abundance in the U.S., but we are losing balance. The pain of hard work, dedication, sacrifice, and commitment are on the avoidance list. Historians will tell us that these trends have occurred before. When proper balance between the two characteristics of pain and pleasure are lost, then all else is lost. Here, too, is another area that needs attention in order to achieve our full measure of deliverance.

CONCLUSION

My commitment to the nontechnologist has been met. MOT has been spelled out in a way that is clear and nontechnical. The nontechnologist can now manage technology and the expert can be managed.

All of the pitfalls have been exposed and solutions prescribed that are practical and workable. The major element needed for success is appropriate leadership. It is time for the metamorphosis of automation-driven management.

It is our sincere hope that in the not-too-distant future, we will have opportunity to reminisce over the success story of winning the Great Economic War of the 1990s. Victory will be sweet, for our children will reap the harvest created by their forefathers. That has been the tradition upon which the American Dream has been founded. That tradition will always be worthwhile and must be preserved at all costs.

Fortunately, the rigors of MOT and economic warfare are not as demanding as other kinds of conflict. Instead of life-threatening obstacles, there are demands for professionalism, tenacious business ability, knowledge, and an appropriate kind of management. The process by which we win can be dignified, challenging, self-developing, and, finally, rewarding in innumerable ways. The avoidance of deliverance will only occur if the prescription outlined in *Harnessing Technology* is avoided. The whole process can be comfortable and uplifting.

The fabric of U.S. society will not fade away or become tattered under this prescription. The foundation for a sound future will be established and tangible evidence of favorable results will become available to our total populace. We will be building a better America in real terms, and we can safely say that all great empires have not fallen.

There is no "hocus pocus" in MOT. Instead, there is simply hard work based on sound knowledge and purpose. It is this same ethic that has been successful for America for two centuries. We still use it for most things; why not MOT? The problem is not in the applicability, but more so the way to apply it. Now with understanding of the process, the path should be clear. The manager will be more apt to rally and take up the cause in our traditional way.

Long-term economic growth in real terms awaits us, for we are now prepared to stop the damaging Doom Boom. It is time to go back to the basics and assure a sound manufacturing base for the future.

With comprehension of AMT through continuous learning, we develop a state of comfort with the technologies that can carry us to WCM status. The strategic linkage must be established to solve the second class citizen dilemma.

The organization shall take on a culture buster motif and expeditiously move to a futuristic mode, while the dissolution of the vendor user problem will eliminate a major obstacle to the implementation of AMT. The test tube crisis will serve as the springboard of awareness to new frontiers and finally MOT for the nontechnologist becomes a practicing reality. The task has been defined and the method prescribed. The avoidance of decadence and the quest for deliverance is no longer out of reach of the automation-driven manager.

It is time to mobilize our resources and take a proactive stance regarding *Harnessing Technology*. Corporate America will flourish and long-term economic growth will be assured. We have now empowered the manager to be master of his own destiny. Although we have treated MOT in a comprehensive way, it is likely with time, that future unforeseen obstacles will arise and challenge our resilience again and again. It only remains to caution the players that the world is by no means a level playing field. Consequently, we must be in a constant state of preparation to step forward, put complacency aside, and overcome the obstacles that most assuredly will threaten us in the future.

Our global competitors prefer to work in quiet ways. They hope that we fail to create the public awareness necessary to rally the populace. They do not encourage us to wipe out the trade deficit, the budget deficit, or turn off the gushing technology stream. Instead, they prefer that we remain fat, happy, unaware, complacent, and bogged down in the quagmire.

That is precisely why *Harnessing Technology* has outlined the formula that opposes these positions. The solution is at last available to the majority. This revelation lies in the concept that the manager need not be an expert to manage the expert. It only remains to enthusiastically embrace this tenet and pursue all that we have painstakingly written here. MOT for the Nontechnologist is the unleashed power, and you are the empowered. May your endeavors be rewarded, for our future is in your hands—the captains of industry.

Bibliography

Auer, Joseph, and Harris, Charles E. *Major Equipment Procurement.* Van Nostrand Reinhold, New York, 1983.

Bessant, J. "The Integration Barrier: Problems in the Implementation of Advanced Manufacturing Technology." *Robotica,* 1985.

Bhalla, Sushil. *The Effective Management of Technology.* Addison Wesley Publishing Company, Reading, Massachusetts, 1987.

Bhote, Keki R. *Supply Management.* American Management Association, New York, 1987.

Brody, H. "Overcoming Barriers to Automation." *High Technology,* May 1985.

Buehler, Vernon M., and Shetty, Krishna A. *Productivity Improvement.* American Management Association, New York, 1981.

Carey, Kevin E. "Factory Automation: Greatly Needed But Slow in Coming. Why?" *Managing Automation,* February 1987.

Casner-Lotto, Gil & Associates. *Successful Training Strategies.* Jossey Bass Inc., San Francisco, 1988.

Chiantella, Nathan A., IBM Corporation. *Management Guide for CIM.* Association of Society of Manufacturing Engineers, Dearborn, Michigan, 1986.

Clark, Kim, Hazio, Robert, and Lorenz Christopher. *The Uneasy Alliance.* Harvard Business School Press, Boston, 1985.

Cribbin, James J. *Leadership: Strategies for Organizational Effectiveness.* Amacom, New York, 1981.

Dertouzos, Michael L., Lester, Richard K., and Solow, Robert M. *Made In America.* MIT Press, Cambridge, Massachusetts, 1989.

Destanik, Robert. *Managing to Keep the Customer.* Jossey Bass Publishers, San Francisco, 1987.

Ferraro, Richard A., Hunley, Terry E., and Shackney, Orry Y. "Banishing Management Barriers to Automation." *Manufacturing Engineering,* January 1988.

Shetty, Y. K., and Buehler, V., Editors. *Productivity and Quality Through Science and Technology.* Greenwood Publishers, Connecticut, 1988.

Graham, Glen A. *Automation Encyclopedia.* Society of Manufacturing Engineers, Dearborn, Michigan, 1988.

Greehalgh, K. "Counterimplementation: Management & Implementation of High Technology Systems." Paper presented at the conference on Economic, Social, Financial & Technical Effects of Automation, Salford University, November 1984.

Gunn, Thomas G. *Manufacturing for Competitive Advantage.* Ballinger Publishing Company, Cambridge, Massachusetts, 1987.

Heizer, Jay, and Render, Barry. *Production and Operations Management.* Allyn & Bacon Inc., Boston, 1988.

Hughes, Katherine. *Corporate Response to Declining Rates of Growth.* Lexington Books, Lexington, Mass., 1982.

Imai, Masaaki. *Kaizen.* Random House, New York, 1986.

Industrial Research Institute. "Position Statement on U.S. Technology Policy Manufacturing Competitive Frontiers (MPC)." Manufacturing Productivity Center Monthly, Chicago, Illinois, November 1988.

Johnston, Wesley J. *Patterns in Industrial Buying Behavior.* Praeger Publishers, New York, 1981.

Kantrow, Alan. *Survival Strategies for American Industry.* Harvard Business Review, John Wiley & Sons, New York, Introduction, 1983.

Lawrence, Paul R., and Dyer, Davis. *Renewing American Industry.* The Free Press, New York, 1983.

Leibson, David E. "Getting on the Productivity Learning Curve." *Productivity and Quality Through Science and Technology,* Shetty, Y. K. and Buehler, V., Editors, Greenwood Publishers, Connecticut, 1981.

Manufacturing Studies Board, Commission on Engineering and Technical Systems, National Research Council, *Towards a New Era in Managing,* National Academy Press, Washington, DC, 1986.

Manufacturing Studies Board. *Human Resource Practices for Implementing Advanced Manufacturing Technology.* National Academy Press, Washington, DC, 1986.

Mitroff, Ian I. *Business Not As Usual.* Jossey Bass Publishers, San Francisco, 1988.

Moore, John. "U.S. Competitiveness and The National Perspective." *Productivity and Quality Through Science and Technology,* Shetty, Y.K. and Buehler, Vernon, Editors, Part III. Greenwood Publishers, Connecticut, 1988.

Morrison, Catherine, McGuire, Patrick E. and Clarke, Mary Ann. *Key to U.S. Competitiveness.* Conference Board Research Report No. 907, The Conference Board Inc., New York, 1988.

Office of Technology Assessment. *Technology and the American Economic Transition.* Congress of the United States, U.S. Government Printing Office, 1988.

Ramo, Simon. *The Business of Science.* Hill & Wang, New York, 1988.

Rosow, Jerome M. and Zager, Robert. *Training the Competitive Edge.* Jossey Bass Publishers, San Francisco, 1988.

Shea, Gordon F. *Practical Ethics,* AMA Publications Division, New York, 1988.

Shetty, Y. K., and Buehler, Vernon, Editors. *Productivity and Quality Through Science and Technology.* Greenwood Publishers, Connecticut, 1988.

Skinner, Wickham. *Managing the Formidable Weapon.* John Wiley & Sons, New York, 1985.

Snyder, Kenton R., and Elliott, Charles S. "Barriers to Factory Automation: What Are They and How Can They Be Surmounted." *Industrial Electronics,* April 1988.

Sommers, Albert T. *The U.S. Economy Demystified.* Lexington Books, Lexington, Massachusetts, 1988.

Steele, Lowell W. *Managing Technology.* McGraw Hill, New York, 1988.

Task Force on Management of Technology. *Management of Technology, The Hidden Competitive Edge.* National Research Council, National Academy Press, Washington, DC, 1987.

Tuleja, Tad. *Beyond the Bottom Line.* Facts on File Publications, New York, 1985.

U.S. Congress, Office of Technology Assessment. *Paying the Bill: Manufacturing and America's Trade Deficit.* OTA-ITE-390. U.S. Government Printing Office, Washington, DC, June 1988.

Walsh, John J. "Vendors Must Change to Meet the User Needs in the 80's." *The Consultant,* Vol. 2, No. 1, January 1984.

Walsh, Susan. "Senate Hearing: Automating Small Manufacturing." *Managing Automation,* February 1988.

Wendt, Henry. *The Corporation of the Future,* Chapter I, "Views from the Top." Rosow, Jerome, Editor, New York, Facts on File, 1985.

Wickham, Penelope. *Insider's Guide to Demographic Know How.* American Demographic Press, Ithaca, New York, 1988.

Index

Index